Lecture Notes

on

Applied

Reservoir

Simulation

Leonard F Koederitz

University of Missouri-Rolla, USA

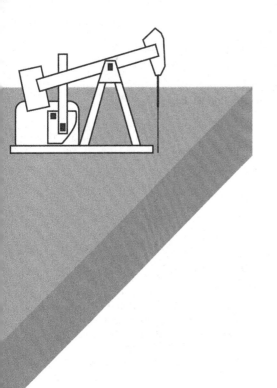

Lecture Notes

on

Applied

Reservoir

Simulation

 **World Scientific**

NEW JERSEY • LONDON • SINGAPORE • BEIJING • SHANGHAI • HONG KONG • TAIPEI • CHENNAI

Published by

World Scientific Publishing Co. Pte. Ltd.

5 Toh Tuck Link, Singapore 596224

USA office: 27 Warren Street, Suite 401-402, Hackensack, NJ 07601

UK office: 57 Shelton Street, Covent Garden, London WC2H 9HE

British Library Cataloguing-in-Publication Data
A catalogue record for this book is available from the British Library.

ISBN-13 978-981-256-198-5
ISBN-10 981-256-198-6

Typeset by Stallion Press
Email: enquiries@stallionpress.com

Printed in Singapore

PREFACE

The concept for *Lecture Notes on Applied Reservoir Simulation* probably began in 1971 while I was working for the (then) Atlantic Richfield Company. Reservoir simulation was a new and innovative concept, taxing computers and engineers to their fullest. Early model developers (of which I was one) were minimal users and as a result, most simulation courses (and eventually, books) dealt to a large extent with simulation theory and development. I was fortunate to be in a position to perform model studies and debug user problems; while many of these problems were actual model errors (especially early on), a fair amount of the discrepancies resulted from a lack of understanding of the simulator capabilities, or inappropriate data manipulation.

On joining the faculty of the University of Missouri-Rolla, I determined that since the majority of petroleum engineering students were undergraduates, they would in all likelihood be model-users, not developers, and in 1975, the first (and minimal) edition of these notes was created. The notes have been updated annually since that time to reflect changes in both simulation concepts and philosophy.

The Society of Petroleum Engineers' publication of *Reservoir Simulation*, Monograph Volume 13 (referred to as M-13 throughout this book) is an excellent collection of simulation examples from various literature sources; however, it is what the name implies (monograph: a treatise on a particular subject) and not necessarily a suitable stand-alone instructional text. Also, note that all of the authors are affiliated with the same company, resulting in somewhat similar viewpoints.

Finally, these lecture notes are as titled, *Lecture Notes on Applied Reservoir Simulation*, not "how to build models" or an in-depth course in reservoir engineering, although portions of each concept must be included in any discussion of reservoir simulation. I have tried to stay with "tried and true" simulation practices and have mentioned new methods which appear to be useful in applied modeling.

L. F. Koederitz
August 2004

CONTENTS

Chapter 1

INTRODUCTION

Reservoir simulation, or modeling, is one of the most powerful techniques currently available to the reservoir engineer. Modeling requires a computer, and compared to most other reservoir calculations, large amounts of data. Basically, the model requires that the field under study be described by a grid system, usually referred to as cells or gridblocks. Each cell must be assigned reservoir properties to describe the reservoir. The simulator will allow us to describe a fully heterogeneous reservoir, to include varied well performance, and to study different recovery mechanisms. Additionally, due to the amount of data

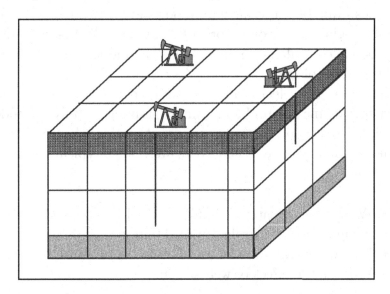

required, we often will reconsider data which had previously been accepted. To make the model run, we perturb the system (usually by producing a well) and move forward in time using selected time intervals (timesteps) as indicated in Fig. 2.5, p. 9, M-13. The main type of results that we gain from a model study are saturation and pressure distributions at various times as shown on p. 8, M-13; quite frequently, these variations will indicate what the primary drive mechanism is at any given point in time.

On the other hand, modeling requires a computer with a fair amount of memory and a great deal of engineering time; you cannot do a model study in an afternoon! It takes time to locate the data, modify it to fit your grid system, enter it and then to actually run the model. Minimum time for a very simple study is a week; average time is probably from 3 to 6 months; large and/or complex studies may encompass years. In short, it takes much more effort on your part to interpret the results of a simulator and as a result, small screening models may be used to evaluate key parameters while larger models would simulate an entire field in detail. As the field is developed and more data becomes available, intermediate models are often developed for specific regions or recovery processes in the field; these models may be called scalable models, but changing the grid presents additional problems.

To be able to decipher what the model is telling you, you must first define the problem. Simply running a study to model a field is not good enough; you must decide ahead of time what questions you are trying to answer. Some typical questions might be:

- What type of pattern should be used for water injection?
- Should a well be drilled in a certain location?
- How would rate acceleration affect the ultimate recovery?
- What is the effect of well spacing?

- Is there flow across lease lines?
- Will the oil rim rise to saturate the gas cap?
- Should gas injection be considered? If so, for how long?
- Should water injection be considered? If so, at what rates?

Once we have decided what questions need to be answered, we can construct the model grid.

1.1. Types of Models

There are five types of models, depending on the grid selected, that may be used (although the first two types are used minimally today):

- One-dimensional horizontal
- One-dimensional vertical
- Areal (two-dimensional)
- Cross-sectional (two-dimensional)
- Three-dimensional

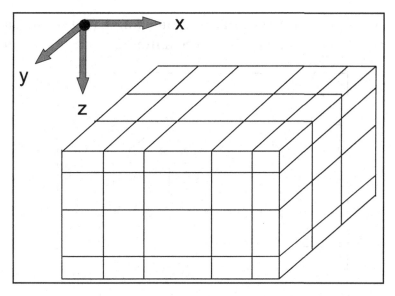

Model grid — Cartesian coordinate system

Various grid systems are illustrated on p. 7 in M-13. Additionally, the coordinate system, number of components (or phases) and treatment of the flow equations yield a large number of simulation possibilities. The most common coordinate system in use is that of cartesian (rectangular) coordinates.

A **one-dimensional (1-D) model** may be used to define a bottom water drive, determine aquifer activity, yield an accurate material balance or as a screening tool prior to a large complex study. Gravity drainage may be simulated using a 1-D vertical model. Sensitivity studies may be conducted and interpreted rapidly using 1-D models; these studies might include the effects of vertical permeability, injection rate, relative permeability, residual oil saturation, reservoir size, etc. This information would be extremely useful in more complex studies. Individual well behavior cannot be modeled using a 1-D model; however, field behavior may be approximated. Trying to match production history of individual wells using a 1-D model is both fruitless and time consuming. 1-D models are seldom used extensively today.

There are two types of **two-dimensional (2-D)** cartesian models; the most common is the **areal model**. Strictly speaking,

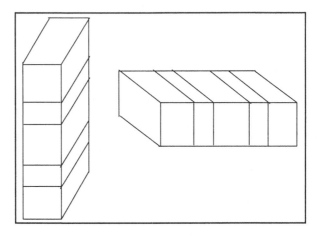

One-dimensional models

an areal model should be used only if there will be very little vertical movement of fluids as in a thin sand; however, the areal model is also employed for thick sands when no great differences in permeability exist (i.e., permeability layering). Dip can be incorporated in an areal model, although water underrunning or gas overriding may not be in its proper perspective if permeability layering exists. The effects of varying well patterns, both in type and spacing may be studied with an areal model.

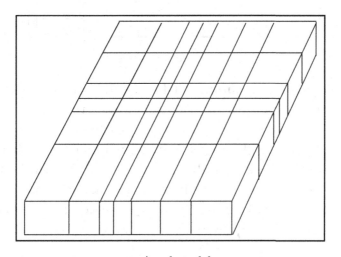

Areal model

The other type of 2-D cartesian model, the **cross-sectional model**, is often used to simulate a slice of a field. It will show vertical and horizontal movement, but is not useful for determining well patterns. Its greatest usage is in determining completion intervals and stratification effects. Usually, when orienting a cross-sectional model (commonly called an X-Z model), the cross-section is taken parallel to the fluid movement (up or down dip). This type of model is used for thick, layered reservoirs, water underrunning, gas segregation, or a series of reservoirs co-mingled in the wellbore.

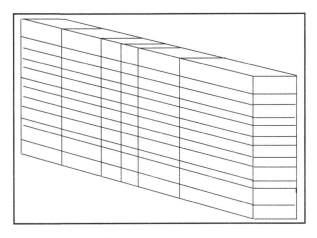

Cross-sectional model

The **three-dimensional (3-D) model** can handle any and all of the previous types of studies; however, the computer time and interpretive engineering time are greatly increased over that required for 2-D models. A 3-D model must be used when fluid migration is expected parallel to the strike of a thick steeply dipping bed (i.e., fluids will flow up dip and across dip). If a typical section of a field cannot be determined for use in a 2-D model, then a 3-D model is required; however, finely modeling

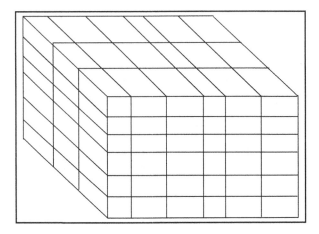

Three-dimensional model

the area of concern and "lumping" the remainder of the field into a few large cells may save considerable time and money as shown in the windowed model (Fig. 3.29, p. 24, M-13). Once again, you must define your problem before you start to model it.

The second type of coordinates employed in simulation is the **radial (R-Z-Θ)** or cylindrical system and may exist in one to three dimensions. Radial systems in two dimensions (R-Z) are sometimes referred to as **coning models** based on their early applications for studying the effects of coning phenomena. They are single well models designed to study individual well effects; additional wells may be included, but they will not exhibit the performance shown in actual production. Coning models are fully implicit in order to handle the rapid saturation changes that occur near the wellbore. Field studies (whole or partial) may also be performed using a cylindrical system, but this application has found limited use. Aquifers may be simulated in radial models by use of a water injection well in the outer block; this technique works well for strong aquifers but may present problems with weaker water drives. Radial models may be used to study coning, shale breaks, well tests, vertical

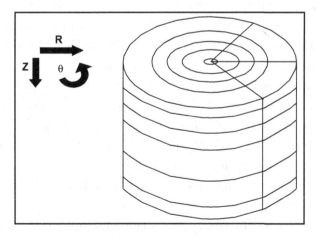

Radial coordinate system

permeability effects, heterogeneity, and to determine maximum producing rates; however, when studying coning, after shut-in, the cone will fall in a simulator without hysteresis; whereas in reality, the cone will not completely drop and imbibition effects will greatly inhibit future production; this concept is discussed in Chaps. 4 and 9 of this book.

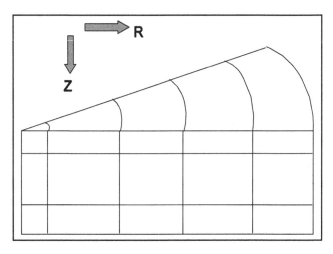

Two-dimensional radial model

Black oil (or **Beta**) **models** consist of three phase flows: oil, gas, and water, although additional gas or aqueous phases may be included to allow differing properties. These models employ standard PVT properties of formation volume factors and solution gas and are the most common type of simulator. PVT properties are covered later in this chapter and in Chap. 3.

Compositional simulators are similar to black oil models as far as dimensions and solution techniques are concerned; here, the similarity ceases, for while volume factors and solution gas effects are employed in a black oil model, a compositional model employs Equations of State (EOS) with fugacity constraints, and uses equilibrium values, densities and several varying components (including non-hydrocarbons). Considerable time is required in the phase package (i.e., matching lab data with simulator requirements) before the actual model can be run. It is reasonable to state that this type of model requires additional expertise to be useful.

Finally, treatment of the model equations yields either an **IMPES (implicit pressure, explicit saturation)** formulation, a **fully implicit** formulation, or some combination thereof. Very simply, an IMPES model is current in pressure and solves for saturations after pressures are known while a fully implicit model solves for both pressures and saturations simultaneously. Rapid saturation changes require fully implicit models. The semi-implicit treatment is a combination which attempts to estimate what saturations will exist at the end of the timestep.

1.2. Data Requirements

Variables required for assignment to each cell (location dependent):

- Length
- Width
- Thickness
- Porosity
- Absolute permeabilities (directional)
- Elevation
- Pressure(s)
- Saturations

Variables required as a function of pressure:

- Solution gas–oil ratio
- Formation volume factors
- Viscosities
- Densities
- Compressibilities

Variables required as a function of saturation:

- Relative permeability
- Capillary pressure

Well data:

- Production (or injection) rate
- Location in grid system
- Production limitations

A similar but more confusing outline of the data required for modeling is on p. 30, M-13.

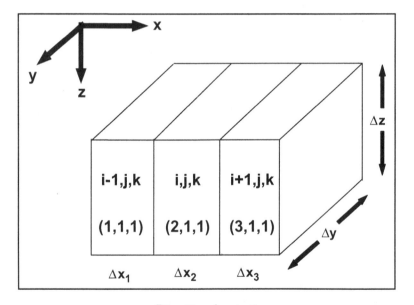

Directional notation

Lengths are normally obtained by superimposing a grid system on a field map and measuring the appropriate distances. These increments are usually denoted using the variable Δx with the subscript "i" referring to the cell location by column (running from left to right). The standard practice of overlaying a grid on a map is used for one-dimensional (both horizontal

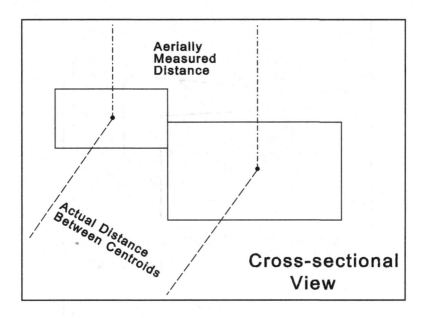

and vertical), areal and three-dimensional models. For dipping reservoirs, the aerial distances will be shorter than the actual distances between the wells. Usually, this discrepancy is not apparent due to the available accuracy of several of the reservoir descriptive parameters, particularly for dip angles of less than 10°; however, the variation may be corrected using pore volume and transmissibility modifiers or as an input option in some simulators. The actual length is $r = x/\cos\Theta$.

Widths are measured in the same manner as lengths and the same discussion applies. Note that the widths in a cross-sectional model need not be constant. Widths are denoted as Δy with a subscript "j" and are sequenced by rows from rear to front (top to bottom in an areal model).

Thickness values are obtained from seismic data, net isopach maps (for areal and 3-D simulations), well records, core analysis and logs (for cross-sectional models). Thicknesses in an areal model may vary with each cell and are denoted as Δz. For layered models the subscript "k" is employed to denote the

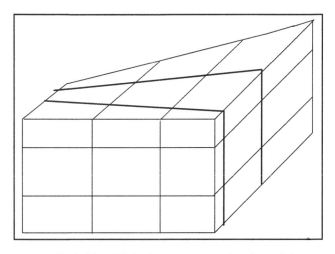

Variable widths in a cross-sectional model

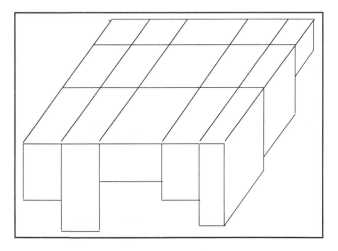

Variable thicknesses in an areal model

layers; they are sequenced from top to bottom. For areal considerations (including 3-D), thickness values may be obtained by superimposing a grid on a net pay isopach. Obviously, thickness values may also be obtained by subtracting the bottom of the formation from the top of formation when these maps are available; at this point, gross pay is known and must then be reduced to net pay. Note that unless a net-to-gross input option is employed, thickness must be a net pay.

When constructing a cross-sectional model using well records and logs, the actual distance between cell centers (centroids) is employed; however, the pore volumes calculated in this instance are in error when (vertical) net pay is used since they are calculated based on (length * width * net pay * porosity). Note that the error introduced tends to compensate for the length error previously discussed.

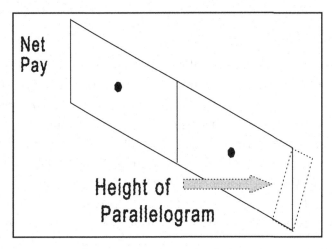

Dip angle effect on thickness

Porosity (ϕ) is a ratio of void space per bulk volume and may be found using logs, laboratory analysis, correlations, and/or isoporosity contour maps. If thicknesses have already been determined, porosity values may be calculated from isovol (ϕh) maps when available.

Total porosity is a measure of total void space to bulk volume whereas **effective porosity** is the ratio of interconnected pore space to bulk volume. For intergranular materials, such as sandstone, the effective porosity may approach the total porosity; however, for highly cemented or vugular materials, such as limestones, large variances may occur between effective and total porosity. In shales, total porosity may approach 40% whereas the effective porosity is usually less than 2%.

Since effective porosity is concerned with the interconnected void spaces, it should be input to simulators. Note that porosity values obtained from logs (Sonic, Density, or Neutron) will approach a total porosity value.

Hydrocarbon porosity is a measure of the pore space occupied by oil and gas to bulk volume and may be defined as

$$\phi_h = \phi(1 - S_w).$$

Porosity is independent of rock grain size but is dependent on the type of packing. A maximum porosity of 47.8% is obtained from cubic packing and a porosity value of 26.0% exists for rhombohedral packing. In general, porosity values for unfractured systems will range from 0 to 30% with the majority of values occurring from some minimum value to 20%. Porosities may be obtained at either reservoir or a fairly low ($\sim$100 psi) pressure in the laboratory, although low pressure values are more commonly reported; log-determined values will be at reservoir

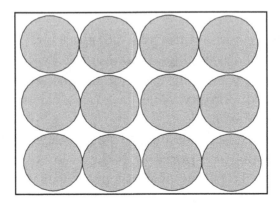

Cubic packing: 47.8%

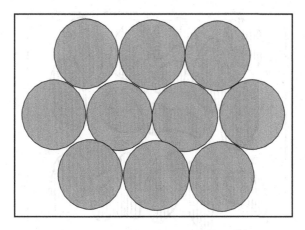

Rhombohedral packing: 26.0%

pressure. The effect of pressure on porosity is

$$\phi_2 = \phi_1 e^{c_f(p_2 - p_1)}$$

which is sometimes written (using a series expansion) as

$$\phi_2 = \phi_1[1 + c_f(p_2 - p_1)].$$

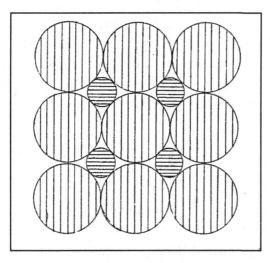

Cubic packing — Two grain sizes: 14%

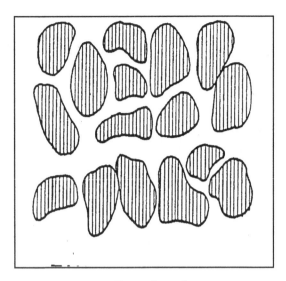

Typical sand

This equation should not be used for extremely soft formations ($\sim$100 microsips); always use the exponential form of the equation. Note that as pressure decreases, porosity decreases due to the overburden effect; however, to convert low pressure lab-measured values to reservoir conditions, the pressure change $(p_2 - p_1)$ must be reversed to $(p_1 - p_2)$. Changes in porosity can account for compaction in highly compressible formations; compaction may or may not be reversible. When averaging porosity values, use a net pay weighted average:

$$\phi_{\text{avg}} = \sum_{i=1}^{n}(\phi_i h_i) \bigg/ \sum_{i=1}^{n} h_i.$$

Additional information concerning porosity may be found on pp. 29–31, M-13.

Absolute permeability (k or k_a) is a measure of the rock capability to transmit fluids. Absolute permeability has units of

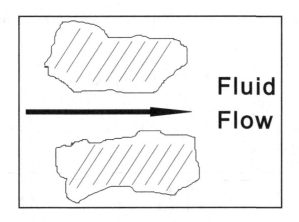

millidarcies (md) and may be obtained from well tests, laboratory analysis, correlations or in rare instances, isoperm maps. Several different techniques are available for analyzing a variety of well tests. Remember that laboratory results apply only to the section of core being analyzed while a well test indicates an average permeability in a region (usually large) surrounding the wellbore.

Also, well test analyses yield effective permeability values and the relationship between effective and absolute permeability is

$$k_e = k_a k_r,$$

where the relative permeability (k_r) is a reduction due to the presence of other fluids, and will be discussed later in this chapter. Comparisons of core data and well test data are shown on p. 35, M-13. Often, permeability will correlate with porosity; some sample correlations of permeability as a function of porosity for core data are shown in Figs. 4.10–4.12, pp. 35–36, M-13.

Three techniques may be used to calculate average permeability values: arithmetic (or parallel), reciprocal (or series or harmonic) or geometric averaging.

For cartesian systems having "nz" layers, the **arithmetic average** is

$$k_{\text{arith}} = \sum_{i=1}^{nz}(k_i h_i) \bigg/ \sum_{i=1}^{nz} h_i$$

which may be used to calculate the horizontal permeability in stratified systems.

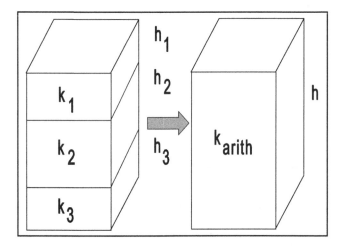

Parallel averaging

The **reciprocal average** for cartesian systems with "nx" columns in series is

$$k_{\text{recip}} = \sum_{i=1}^{nx} L_i \bigg/ \sum_{i=1}^{nx} \frac{L_i}{k_i}$$

which is represented as shown.

A third technique sometimes employed in averaging permeabilities for randomly distributed data is the **geometric average**

$$k_{\text{geo}} = \exp\left[\left(\sum_{i=1}^{n} h_i \ln k_i\right)\bigg/ \sum_{i=1}^{n} h_i\right]$$

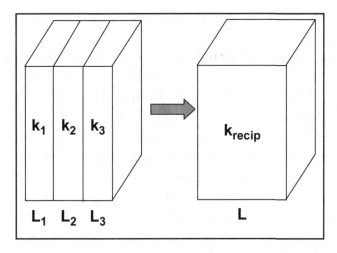

Series averaging

or for "n" evenly spaced intervals,

$$k_{\text{geo}} = \sqrt[n]{\prod_{i=1}^{n} k_i}.$$

Note that the reciprocal average favors smaller values and that the geometric average falls somewhere between reciprocal and arithmetic averaging results.

Additionally, permeabilities may have directional trends (**anisotropy**); for example, in an areal model, the North–South permeability may be greater than the East–West permeability. In standard cartesian gridding, there may only be two areal permeabilities which must be orthogonal and as such, the grid must be aligned with any directional trends. In cross-sectional and 3-D models, **vertical permeabilities** are required; for example, a sealing shale in a cross-sectional model would have a vertical permeability of zero. Quite frequently, a value of one-tenth of horizontal permeability is used for vertical permeability (note that this method is not necessarily recommended, only mentioned). Both vertical and areal permeability variations may be determined by well tests. M-13 discusses absolute permeability on pp. 31–38.

Elevations (or **depths**) for areal and 3-D models are usually obtained from structure maps which have been constructed based on data obtained during drilling and logging as well as other geological information as shown in Fig. 8.9, p. 96, M-13. The variable used to denote elevations is usually D or h; this may prove confusing, since h is usually used for net pay (which is Δz in most simulators). The simulator requires the elevation at the centroid of each cell so that top of formation or bottom of formation maps should be adjusted to the center of the cells. Many simulators will accept top of sand data and adjust it by one-half of the net pay. Elevations may be referenced from any convenient (and consistent) location: subsea, subsurface (when horizontal), kelly bushing, marker sand, or even top or center of formation. In most models, the directional notation is that down (from the reference elevation) is positive and up is negative. For smoothly dipping reservoirs, the rate of dip (ft/mile) may be calculated as

$$5280 \tan \Theta,$$

where Θ is the dip angle; often, this calculation is shown as $5280 \sin \Theta$ and for dip angles of $10°$ or less, the sine and tangent are numerically similar.

In constructing a cross-sectional model from well records and logs, the procedure is similar to that described using structure maps. For layered models (cross-sectional and 3-D), elevations may be required for every cell in every layer; when no gross discontinuities exist, the top layer elevations may be adjusted by averaging the pay zones; however, when the actual reservoir zones are separated by non-productive rock, elevations must be determined for each cell.

Pressures are required for each cell in a simulator and may be input on a per cell basis; however, if the simulation begins at

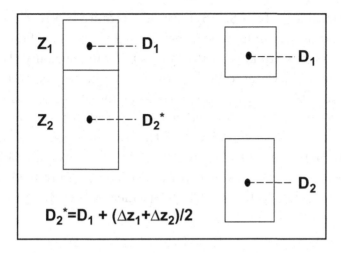

Layer elevation calculations

equilibrium conditions, it is much easier to use a pressure at a known datum and calculate pressures for all cells using a density gradient adjustment

$$P = P_{\text{datum}} + \frac{\rho\,\Delta D}{144},$$

where

P = pressure in cell, psia
P_{datum} = datum pressure, psia
ΔD = change in elevation, ft (+ is down)
ρ = fluid density, lb/ft^3.

Additionally, in multiphase flow, a pressure for each phase (oil, gas and water) must be calculated. The pressure in the water phase is related to the oil pressure by the capillary pressure

$$P_w = P_o - P_{c_{wo}}$$

and the pressure in the gas phase is related to the oil pressure by

$$P_g = P_o + P_{c_{go}}.$$

Saturations (S_o, S_w, S_g) are also required for each cell; as with pressures, they may be directly assigned to cells; however, if the saturations are known at any given datum (usually the gas–oil contact and water–oil contact), they may be determined at equilibrium based on capillary pressures for each cell. For example, to determine the oil and water saturations 10 feet above the water–oil contact (defined in this example as 100% water) for a 50 lb/ft^3 oil and a 65 lb/ft^3 water, the water–oil capillary pressure, at the contact, is 0 psi (since no oil is present). If the pressure at the WOC is 3000 psi (which is a water pressure), then

$$P_o = P_w + P_{c_{wo}}$$
$$= 3000 + 0$$
$$= 3000 \text{ psi}.$$

At a point 10 feet above the WOC, the oil pressure is

$$P_o = P_{o\,\text{datum}} + \rho_o \Delta D / 144$$
$$= 3000 + (50)(-10/144)$$
$$= 3000 - 3.5$$
$$= 2996.5 \text{ psi}$$

and the water pressure is

$$P_w = P_{w\,\text{datum}} + \rho_w \Delta D / 144$$
$$= 3000 + (65)(-10/144)$$
$$= 3000 - 4.5$$
$$= 2995.5 \text{ psi}$$

and the capillary pressure 10 feet above the WOC is

$$P_{c_{wo}} = P_o - P_w$$
$$= 2996.5 - 2995.5$$
$$= 1.0 \text{ psi}$$

so the water saturation at this point corresponds to the value which exists at a capillary pressure of 1 psi. This same technique is explained in Sec. 4.7.1 on p. 41, M-13 and will make a lot more sense after the discussion on capillary pressure at the end of this chapter.

Solution gas–oil ratio (R_s) or dissolved gas is required as a function of pressure and based on the pressure in each cell, the amount of solution gas will be calculated for each cell. It may have units of either SCF of solution gas per STB oil, or MCF solution gas per STB oil; in the former case, the values should be between 50 and 1400 SCF/STB with the majority of fields falling between 200 and 1000 over reasonable pressure ranges. Obviously, for units of MCF/STB, the variations are 0.05 to 1.4, etc. Quite frequently, dissolved gas values are given without units and it is necessary to determine the appropriate units. When plotted as a function of pressure, solution gas remains constant above the bubble point and decreases with decreasing pressure below the bubble point as gas is released from solution to become free gas. Although curvature exists below the bubble point, a large number of solution gas samples exhibit a markedly linear relationship, and a reasonable first-guess can

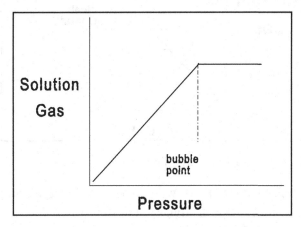

Solution gas plot

often be obtained by using the bubble point value and a dead-oil value of zero at atmospheric pressure.

Two types of liberation processes may be used to measure solution gas: flash and differential. In a **flash liberation process**, gas which is released from solution remains in contact with the oil (a constant composition process) whereas in **differential liberation**, the free gas is removed while maintaining pressure. Flow in reservoirs with any appreciable vertical permeability will approximate a differential process while tubing, surface equipment and reservoirs having continuous shales approach a flash process. Laboratory analyses usually give pressure-dependent differential values of solution gas and a bubble point flash value; pressure-dependent flash values may be calculated using

$$R_{s_{\text{flash}}} = R_{s_{\text{differential}}} \frac{R_{sbp_{\text{flash}}}}{R_{sbp_{\text{differential}}}}.$$

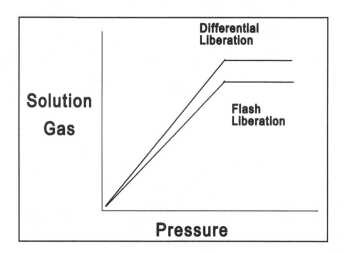

Flash and differential liberation

Most simulation studies will have solution gas values available from a laboratory analysis; however, for some preliminary studies, it may be necessary to estimate dissolved gas using correlations.

Solution gas–water ratio or the dissolved gas in water is required in some models. While the same concept as for dissolved gas in oil applies, the amount of gas soluble in most aquifers is significantly less, ranging from 4 to 20 SCF/STB; R_{sw} is the variable used to denote dissolved gas in water. In general, for oil and gas simulations, omitting the effects of R_{sw} causes no visible change in the results.

Oil formation volume factors (B_o) relate a reservoir volume of oil to a surface volume. The reservoir volume includes dissolved gas whereas the surface volume does not. The oil formation volume factor has units of RVB/STB. A reasonable range for the oil formation volume factor is from 1.05 to 1.40 RVB/STB. Note that the oil formation volume factor includes any dissolved gas; very simply, dissolved gas is considered as part of the oil. Below bubble point pressure, a decrease in pressure results in a decrease in B_o due to the fact that dissolved gas is released from the oil yielding a lesser volume at the lower pressure.

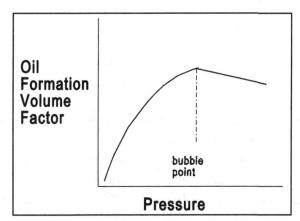

Oil formation volume factor plot

Above the bubble point (in an undersaturated condition), a decrease in pressure releases no solution gas and when relieving the pressure on a fixed volume, expansion occurs, and the oil

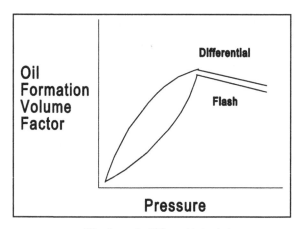

Flash and differential plot

formation volume factor increases (slightly) with a decrease in pressure until the bubble point is reached.

Both flash and differential liberation techniques are used in the laboratory for determining oil formation volume factors and the discussion given for solution gas also applies to the oil formation volume factor. Flash values of the oil formation volume factor may be determined from

$$B_{o_{\text{flash}}} = B_{o_{\text{differential}}} \frac{B_{obp_{\text{flash}}}}{B_{obp_{\text{differential}}}}.$$

Use of flash data may cause severe timestep limitations when going through the bubble point.

The oil formation volume factor is usually a gentle curve up to the bubble point and over a limited pressure range is a fairly straight line above the bubble point. Above the bubble point,

$$B_o = B_{obp}\, e^{-c_o(P - P_{bp})}$$

which is often shown using a power series expansion as

$$B_o = B_{obp}[1 - c_o(P - P_{bp})],$$

where

B_o = oil formation volume factor, RVB/STB (above bubble point)

B_{obp} = oil formation volume factor, RVB/STB (at bubble point)

c_o = undersaturated oil compressibility, psi

P = reservoir pressure, psia

P_{bp} = bubble point pressure, psia.

For highly undersaturated reservoirs, use the exponential form of the oil formation volume factor equation. Note that the oil formation volume factor above the bubble point must always be less than the bubble point value.

Gas formation volume factor (B_g) is a function of pressure; unfortunately, several different units may be applied to the gas formation volume factor: RCF/SCF, RVB/SCF, or RVB/MCF. Since many flow rates are measured in MCF/day and the combination of rate times volume factor is desired, and due to the fact that the values are in the range of 0.1 (at high pressures) to 35 (at low pressures), RVB/MCF is a preferred set of units. For most reservoir pressures encountered, B_g will

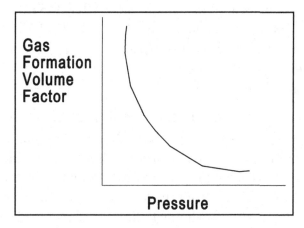

Gas formation volume factor plot

be between 0.2 and 1.5 RVB/MCF. The gas formation volume factor is readily calculated from

$$B_g = \frac{5.035z(T + 460)}{P},$$

where

B_g = gas formation volume factor, RVB/MCF
z = gas deviation factor
T = reservoir temperature, F
P = reservoir pressure, psia.

The gas formation volume factor increases with decreasing pressure due to expansion. Values of the gas deviation factor (z-factor) may be obtained from laboratory analysis of gas samples or correlations such as the z-factor chart by Standing and Katz or the resultant equations of Yarborough and Hall, or others.

Water formation volume factors (B_w) are required as a function of pressure although many simulators employ a value at a base pressure and correct it using

$$B_w = B_{wb}e^{-c_w(P-P_b)} \approx B_{wb}[1 - c_w(P - P_b)],$$

where

B_w = water formation volume factor, RVB/STB
B_{wb} = water formation volume factor at P_b, RVB/STB
c_w = water compressibility, /psi
P = reservoir pressure, psi
P_b = base pressure, psi.

Water formation volume factors are usually very close to 1.0, ranging from 1.0 to 1.05 RVB/STB. Due to the small amount of gas dissolved in water, the formation volume factor will increase slightly with decreasing pressure. Water formation volume factor data is seldom available from the lab and correlations are usually employed; this is due to the fact that the slight deviation from 1.0 usually does not warrant the expenditure for a lab analysis.

Oil viscosity (μ_o) is a measure of the molecular interaction (the intertwining of hydrocarbon chains) and is required as a function of pressure in simulators; standard units are centipoise (cp). Frequently, it is available from laboratory analyses, either at a base pressure or reservoir pressures. If unavailable, it may be estimated (or corrected from base pressure to reservoir conditions) from correlations. Oil viscosity increases with decreasing pressure at saturated conditions (below the bubble point) due to the release of solution gas (small molecules compared to the oil). Above the bubble point, a decrease in pressure yields a decrease in oil viscosity because the molecules are not forced as close together as at the higher pressure.

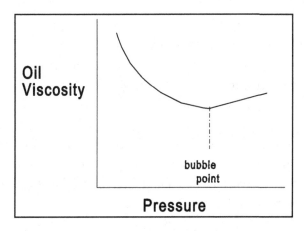

Oil viscosity plot

Gas viscosity (μ_g) is primarily a function of pressure; when measured in the laboratory, it may be reported at a base pressure (usually atmospheric) or at reservoir pressures. As pressure decreases, gas viscosity decreases. A reasonable range of gas viscosity values is from 0.01 to 0.04 cp with higher values at pressures in excess of 10,000 psi. When unavailable as laboratory data, gas viscosities may be estimated using correlations.

Water viscosity (μ_w) is seldom input to simulators at varying pressures due to the fact that it is somewhat independent

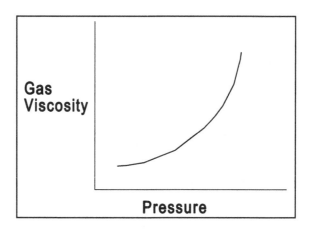

Gas viscosity plot

of pressure being primarily a function of temperature and to a lesser degree, a function of salinity. Sometimes a base pressure and reservoir temperature value is available from lab analysis when required; if not, a correlation may be employed. A normal range for water viscosities at reservoir temperatures is from 0.3 to 0.8 cp.

Oil density (ρ_o) is almost always reported in terms of a stock tank gravity (which is a dead oil); most simulators adjust this value to reservoir conditions using the following relationship below the bubble point

$$\rho_o = \frac{\rho_{oST} + 13.56\, gg\, R_s}{B_o},$$

where

ρ_o = oil density, lb/ft^3

ρ_{oST} = stock tank oil density, lb/ft^3

gg = gas gravity

R_s = dissolved gas, MCF/STB

B_o = oil formation volume factor, RVB/STB.

Above the bubble point, R_{sBP} is used in place of R_s.

Oil densities are normally reported as API gravities and the relationship between API gravity and density in lb/ft^3 is

$$\rho_{oST} = \frac{(62.4)(141.5)}{131.5 + API} = \frac{8829.6}{131.5 + API}.$$

Note that the oil densities will be used to determine the pressure gradients for initialization in the simulator using $\rho_o/144$ to obtain the gradient in psi/ft. A normal range of API gravities is from 45° to 10° corresponding to densities of 50.0 and 62.4 in lb/ft^3 respectively.

Gas density (ρ_g) is usually input as a gas gravity (gg or γ_g) or in units of lb/MCF. The relationship between these two quantities at standard conditions is

$$\rho_{gST} = \frac{(28.9)(14.7)(1000)\, gg}{(10.73)(460 + 60)} = 76.14\, gg,$$

where

$$\rho_{gST} = \text{gas density, lb/MCF}$$
$$gg = \text{gas gravity}$$

and gas densities are generated from

$$\rho_g = \frac{1000\rho_{gST}}{5.615 B_g},$$

where

$$B_g = \text{gas formation volume factor, RVB/MCF}$$

and density gradients are calculated with $\rho_g/144000$ in psi/ft. A normal range for gas gravities is from 0.6 to 1.2 which corresponds to values of 45.7 to 91.4 in lb/MCF.

Water density (ρ_w) is required as either a density in lb/ft^3 or as a specific gravity (γ_w). The relationship between the two is

$$\rho_w = 62.4\,\gamma_w$$

and due to salinity the water density at standard conditions may be estimated from

$$\rho_{wST} = 62.4 + 0.465S,$$

where

$$S = \text{salinity}, \%.$$

Finally, the standard density may be corrected to reservoir conditions using

$$\rho_w = \frac{\rho_{wST}}{B_w}$$

and the gradient calculated from $\rho_w/144$. Most oilfield waters have densities slightly greater than $62.4\ \text{lb}/\text{ft}^3$.

Oil compressibility (c_o) may be defined either above or below the bubble point; however, the only value(s) required in simulators are for undersaturated conditions where the compressibility is used to adjust the oil formation volume factor from bubble point conditions, using either

$$B_o = B_{obp}\, e^{-c_o(P-P_{bp})}$$

or

$$B_o = B_{obp}[1 - c_o(P - P_{bp})]$$

as shown earlier. Oil compressibility may be measured in the laboratory or obtained from correlations. Standard units for oil compressibility are /psi which yields oil compressibility values ranging from 6×10^{-6} to 20×10^{-6}; a more recent unit is the microsip which is 10^6 times greater.

Water compressibility (c_w) is almost always obtained from correlations. For undersaturated conditions, it is usually a number close to 3×10^{-6}/psi (or 3 microsips) at reservoir conditions.

Formation compressibility (c_f), sometimes mistakenly referred to as rock compressibility, is primarily a measure of the pore volume compression of the formation. Data are seldom available and correlations are often employed. The usual range of formation compressibilities (for hard reservoir rocks) is from 3×10^{-6} to 8×10^{-6}/psi although some limestones may exhibit higher values at low porosity.

Relative permeability (k_r) is a reduction in flow capability due to the presence of another fluid and is based on

- pore geometry
- wettability
- fluid distribution
- saturation history.

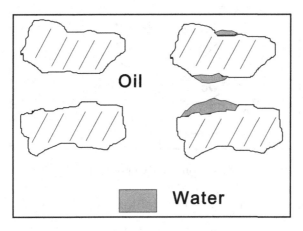

Relative permeability effect

Relative permeability is dimensionless and is used to determine the effective permeability for flow as follows:

$$k_e = k_a k_r.$$

Relative permeability data are entered in models as functions of saturation and may be obtained from laboratory measurements, field data, correlations, or simulator results of a similar formation. Whether appropriate or not, it is usually the first data to be modified in a model study. The simplest concept in relative permeability is that of two-phase flow. For oil reservoirs, the combinations are water–oil and liquid–gas (usually thought of as oil–gas); for gas reservoirs, gas–water applies; and for condensate reservoirs, gas–liquid.

Water–oil relative permeability is usually plotted as a function of water saturation. At the critical (or connate) water saturation (S_{wc}), the water relative permeability is zero

$$k_{rw} = 0$$

and the oil relative permeability with respect to water (or, in

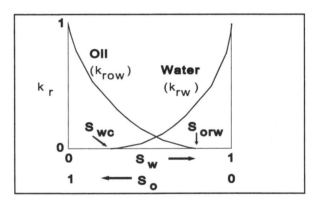

Water–oil relative permeability

the presence of water) is some value less than one

$$k_{row} < 1.0.$$

At this point, only oil can flow and the capability of the oil to flow is reduced by the presence of critical water. Note that data to the left of the critical water saturation is useless (unless the critical water becomes mobile). As water saturation increases, the water relative permeability increases and the oil permeability (with respect to water) decreases. For the oil reservoir proper, a maximum water saturation is reached at the residual oil saturation (S_{orw}); however, since models use an average saturation within each cell, oil saturation values of less than residual oil (in a cell) should be correctly entered. Also, the end point value of $k_{rw} = 1$ at $S_w = 1$ would be required when an aquifer is being included in the simulation study.

Wettability is a measurement of the ability of a fluid to coat the rock surface. Classical definitions of wettability are based on the contact angle of water surrounded by oil and are defined as

$\Theta < 90°$ water–wet

$\Theta > 90°$ oil–wet

$\Theta = 90°$ intermediate or mixed wettability.

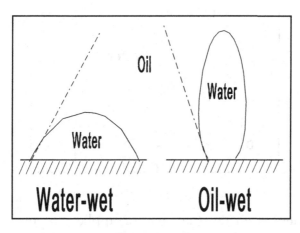

Contact angles

A variation of up to $\pm 20°$ is usually considered in defining intermediate wettability. Contact angle measurements are difficult to perform under reservoir conditions.

Unfortunately, there is a second definition of water-oil relative permeability currently in use, known as **normalized relative permeability**. This method defines the oil relative permeability at critical water as having a value of 1 and defines absolute permeability as the effective permeability with critical water present. In either case, the effective permeabilities will be identical. These values of relative permeability may be corrected to standard values by

$$k_{r\text{STD}} = k_{r\text{NORM}} \frac{k_{a\text{NORM}}}{k_{a\text{STD}}}$$

where

$$k_{a\text{NORM}} = k_{eo} \quad \text{at } S_{wc}.$$

Note that at high water saturations, k_{rw} may exceed a value of one for normalized relative permeability values, particularly for oil-wet systems.

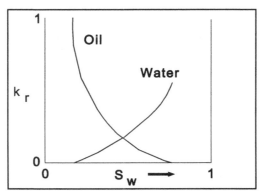

Normalized water–oil relative permeability

Some heuristic rules that may be applied to normalized oil–water relative permeability are shown in the following table; these rules were originated by Craig and subsequently modified by Mohamad Ibrahim and Koederitz (SPE 65631) which would result in the following types of relative permeability plots based on wettability.

Rock Wettability	S_{wc}	S_w at which k_{rw}^* and k_{row}^* are equal	k_{rw}^* at $S_w = 100 - S_{orw}$ (fraction)
Strongly Water-Wet:	$\geq 15\%$	$\geq 45\%$	≤ 0.07
Water-Wet:	$\geq 10\%$	$\geq 45\%$	$0.07 < k_{rw}^* \leq 0.3$
Oil-Wet:	$\leq 15\%$	$\leq 55\%$	≥ 0.5
Intermediate: (Mixed-Wet)	$\geq 10\%$	$45\% \leq S_w \leq 55\%$ OR	> 0.3
	$\leq 15\%$	$45\% \leq S_w \leq 55\%$	< 0.5

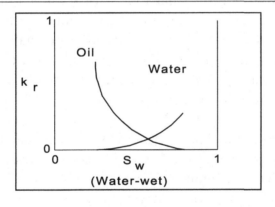

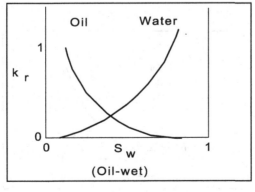

Wettability effect on relative permeability

Gas–oil relative permeability, or gas–liquid relative permeability, is similar in concept to water–oil relative permeability. The preferred relative permeability values are those taken with critical water present. As free gas saturation increases, the oil relative permeability with respect to gas (k_{rog}) decreases until the residual oil saturation with respect to gas (S_{org}) is reached; however, until the critical gas saturation (S_{gc}) is reached, the gas relative permeability is zero ($k_{rg} = 0$). The critical gas saturation is the point at which the gas bubbles become large enough to break through the oil and away from the rock surface. As gas saturation increases, the gas relative permeability increases and theoretically reaches a value of unity at 100% gas. In reality, both in the reservoir and in a simulator, critical water will always be present, so data to the right of the S_{wc} value are meaningless. Incidentally, usually $S_{org} < S_{orw}$. Also, note that if $k_{rog} = 1$ at $S_g = 0$, all values of k_{rog} should be multiplied (reduced) by k_{row} at S_{wc}, but k_{rg} values should **not** be adjusted. A minimal discussion of relative permeability is on pp. 38–40, M-13.

Capillary pressure (P_c) data is required in simulators to determine the initial fluid distributions and to calculate the pressures of oil, gas and water. It is the difference in pressure between two fluids due to a limited contact environment. This data is required as a function of saturations and may be obtained from

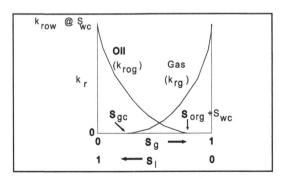

Gas–oil relative permeability

laboratory measurements, correlations or estimated to yield the desired fluid distributions. When laboratory measurements are used, they must be corrected to reservoir conditions

$$P_{cr} = P_{cL}\frac{\sigma_r}{\sigma_L},$$

where

P_{cr} = capillary pressure at reservoir conditions, psi

P_{cL} = capillary pressure at lab conditions, psi

σ_r = interfacial tension of reservoir fluids, dynes/cm

σ_L = interfacial tension of lab fluids, dynes/cm.

When fluid distributions are known at various depths (from core analysis or logging techniques), capillary pressures may be estimated from

$$P_c = \frac{H\Delta\rho}{144},$$

where

P_c = capillary pressure, psi

H = height of transition zone above denser fluid, ft

$\Delta\rho$ = difference in density between two fluids, lb/ft^3 (a positive number).

With rare exceptions (high capillary ranges), capillary pressures have minimal effects once the reservoir is produced.

Water–oil capillary pressure may be determined from either of the two techniques previously examined. It ranges from 0 psi at 100% water to a maximum value of between 5 and 25 psi (usually) at critical water. For extremely homogeneous formations (and from lab data), the (imbibition) curve is as shown; however, for heterogeneous reservoirs, several different sets of the previous curve will ultimately tend toward a straight line; this same linearity occurs due to gridblock effects as shown in Fig. 3.22, p. 21, M-13 and as we will see later, due to fluid segregation.

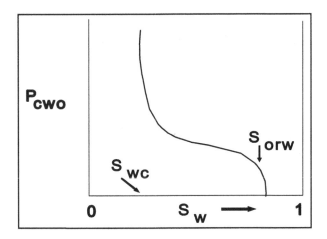

Homogeneous formation capillary pressure plot

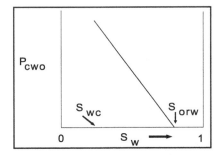

Heterogeneous formation capillary pressure plot

Another correlating factor for water–oil capillary pressure is the *J*-**function** from which capillary pressures may be calculated using

$$P_c = 4.619 J(S_w)\sigma\sqrt{\frac{\phi}{k}},$$

where

P_c = capillary pressure, psi
$J(S_w)$ = *J*-function value at S_w
σ = interfacial tension, dynes/cm
ϕ = porosity, fraction
k = permeability, md

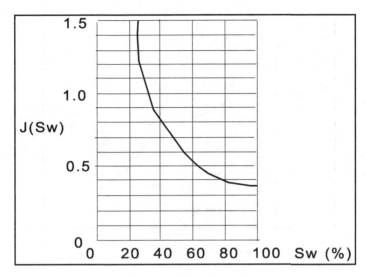

Sample J-function plot

and the *J*-function values are selected at varying water saturations from an appropriate correlation as shown. Allowing input of *J*-functions in place of capillary pressure data would result in capillary pressure tables that vary with the porosity and permeability values assigned to each cell; additionally, varying the interfacial tension with pressures in the PVT table would allow a capillary pressure variation with pressure.

Gas–oil capillary pressure is usually determined by laboratory air–oil data or by estimating the capillary values based on the height of the transition zone. When using the transition zone approach, the gas density may be calculated from

$$\rho_g = 13.56 \frac{gg}{B_g},$$

where

ρ_g = gas density, lb/ft^3
gg = gas gravity
B_g = gas formation volume factor, RVB/MCF.

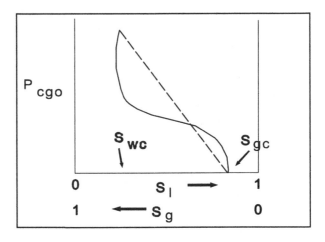

Gas–oil capillary pressure plot

Since most gas–oil transition zones are short, a reasonable range of gas–oil (or gas–liquid) capillary values is from 0 psi at all liquid (or no free gas) to a maximum value of between 2 and 10 psi (usually) at critical water (maximum gas saturation in a gas cap). The discussion concerning linearity (under water–oil capillary pressure) is also applicable to gas–oil data.

Production or **injection rates** are required for each well to be modeled. For liquids, the rate is usually in STB/day and for gas, MCF/day. For producing wells, only one phase production should be specified and that phase is usually the predominant phase. For example, an oil well would specify oil production, and the appropriate gas and water producing rates would be calculated by the model. This data is normally obtained from well files; although the data will vary with time, it is acceptable to use an average rate over a given period of time as long as no drastic rate fluctuation has occurred (see Fig. 7.2, p. 75, M-13). In many cases, a reasonable assumption is to adjust rates when a variation by a factor of 2 occurs. Average rates should be calculated based on production during the time period

$$q_o = \frac{\Delta N_p}{\Delta t},$$

where

q_o = oil rate, STB/day

ΔN_p = oil production during time Δt, days

Δt = production time, days.

Well locations in the grid system are also required as shown in Fig. 5.1(b), p. 45, M-13. Remember the cells are numbered from left to right in the x-direction (the "i" index location), rear to front in the y-direction (the "j" index location), and top to bottom in the z-direction (the "k" index location). In general, for areal and 3-D models, a well should be centered in a cell whenever possible. Vertical layers should correspond to completion intervals in cross-sectional and 3-D models as shown in Fig. 7.5, p. 80, M-13.

Production limitations may be imposed on wells. Some of these may be bottom-hole pressures, skin factors, maximum GOR or WOR limits, total field limitations, coning effects and abandonment conditions (rate, GOR, WOR). Although called production limits, many of these conditions may also be applied to injectors. The end result of all of these stipulations is to adjust the rate data somewhat automatically. A brief discussion of well management may be found in Sec. 2.6 (p. 11), of M-13 and a more complete discussion of well options is in Chap. 7 of this book and Chap. 7 (pp. 74–86) of M-13.

Problems — Chapter 1

1. *Model selection.* The reservoir shown below is sealed by pinchouts. It is undersaturated (above bubble point pressure) and you would expect a depletion drive mechanism from the data available.

It has been proposed to convert one of the four present producers to an injection well to maintain pressure above the bubble point. You have been told to evaluate the reservoir engineering aspects of the proposal.

What tools would you use (type of model, simple calculations, etc.) and what would you expect to learn from each? Which tool would you use first? Which last? How would you grid the field areally for a study?

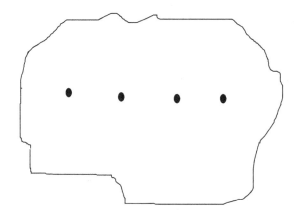

2. *Length calculations.*   In an areal model (using aerial well locations and grids), two cells will be one mile apart (5,280 ft). The reservoir dips continuously at an angle of 6°. What is the

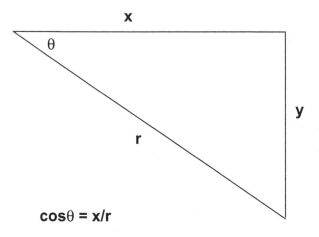

true length between the two cells and how much error have we introduced by using the areal map?

3. *Porosity.* A reservoir is discovered at 5,000 psig having a log-measured porosity of 20%. If the abandonment pressure is 1,000 psig, what value of porosity will exist at abandonment? The formation compressibility is 3.6 microsips (3.6×10^{-6}/psi).

If the porosity of 20% had been measured in a core analysis laboratory at a pressure of 100 psig, what would the value of the original reservoir porosity be?

4. *Permeability averaging.* Calculate the horizontal and vertical permeabilities for the reservoir shown; also, calculate the geometric-average permeability.

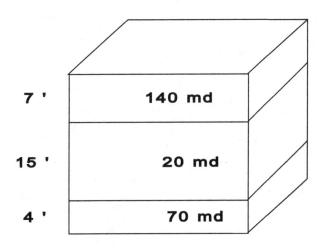

5. *Dip angle.* Calculate the change in elevation over 3 miles for a reservoir having an 8° dip angle. Also determine the resulting additional pressure that would exist at this depth (use a brine hydraulic gradient of 0.466 psi/ft).

6. *Pressure gradients.* The pressure at the gas–oil contact is 2,200 psia in a reservoir; the contact is very sharp and no transition zone exists. What pressure would occur 40′ below the GOC? The oil has a reservoir density of 52.1 lb/ft³.

7. *Laboratory-determined capillary pressure.* Calculate the height of a water–oil transition zone in a reservoir having a critical water saturation of 35%; the laboratory air–water capillarity at the critical water saturation is 18 psi and the air–water interfacial tension is 72 dynes/cm. The stock tank density of the crude is 35° API and it has a water–oil interfacial tension of 24 dynes/cm at reservoir conditions. The water specific gravity is 1.09 and the formation volume factors for water and oil are 1.02 and 1.24 RVB/STB, respectively. The solution gas is 540 and the dissolved gas gravity is 0.7.

8. *Estimation of capillary pressures using the J-function.* Estimate, using the J-function, the capillary pressures for a

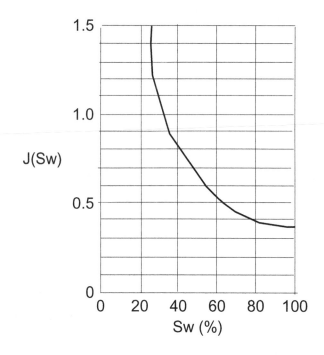

reservoir having a critical water saturation of 22% and an oil–water interfacial tension of 27 dynes/cm. The oil has a gravity of 28° API and the water has a specific gravity of 1.1. The reservoir averages 15% porosity and has a 130 md permeability. Calculate the capillary pressures at saturations of 22, 25, 30, 50, 80, 100%.

9. *Gas–oil capillary pressure.* Calculate the capillary pressure at the top of a GOC consisting of a 5′ transition zone. Oil density is 52 lb/ft^3 at stock tank conditions (38° API) and the gas gravity is 0.75. Formation volume factors for oil and gas are 1.20 RVB/STB and 0.80 RVB/MCF. Solution gas is 0.7.

Chapter 2

THEORETICAL DEVELOPMENT

2.1. Flow Equations

Total Differential

$$Q = f(x, y, z)$$

$$dQ = \frac{\partial Q}{\partial x}dx + \frac{\partial Q}{\partial y}dy + \frac{\partial Q}{\partial z}dz$$

where $(\partial Q/\partial x)$ represents the change in Q with respect to x.

Addition

$$\frac{\partial(A + B)}{\partial x} = \frac{\partial A}{\partial x} + \frac{\partial B}{\partial x}.$$

Multiplication

$$\frac{\partial(AB)}{\partial x} = A\frac{\partial B}{\partial x} + B\frac{\partial A}{\partial x}.$$

Constant

$$\frac{\partial(1)}{\partial x} = 0.$$

Reciprocal

$$\frac{\partial\left(\frac{1}{x}\right)}{\partial x} = \frac{-1}{x^2}.$$

Chain Rule

$$\frac{\partial A}{\partial x} = \left(\frac{\partial A}{\partial y}\right)\left(\frac{\partial y}{\partial x}\right).$$

Finite-Difference Approximation

$$\frac{\partial A}{\partial x} \approx \frac{A_2 - A_1}{x_2 - x_1}.$$

Simplified Theoretical Development of Three-Phase, Three-Dimensional Flow Equations

Based on the continuity equation, we may write a general expression for the mass balance of a flowing system as

$$\text{flow in} - \text{flow out} - \text{production} = \text{accumulation}, \quad (1)$$

where

$$\text{flow in} = Q_x + Q_y + Q_z \quad (2)$$

and

$$\text{flow out} = \text{flow in} + \Delta x \frac{\partial Q_x}{\partial x} + \Delta y \frac{\partial Q_y}{\partial y} + \Delta z \frac{\partial Q_z}{\partial z}. \quad (3)$$

For oil and water phases, we may substitute Darcy's Law for the velocity terms

$$Q_x = -0.00113 \frac{k\, k_r}{\mu B} A_x \frac{\partial \Phi}{\partial x}, \quad (4)$$

where the cross-sectional area open to flow is defined as

$$A_x = \Delta y \Delta z \quad (5)$$

Substituting Eq. (5) into (4) and then, (4) into (3) yields (with minor rearrangement)

flow in − flow out

$$= 0.00113 \Delta x \Delta y \Delta z \left[\frac{\partial}{\partial x} \left(\frac{k\, k_r}{\mu B} \frac{\partial \Phi}{\partial x} \right) + \frac{\partial}{\partial y} \left(\frac{k\, k_r}{\mu B} \frac{\partial \Phi}{\partial y} \right) \right.$$

$$\left. + \frac{\partial}{\partial z} \left(\frac{k\, k_r}{\mu B} \frac{\partial \Phi}{\partial z} \right) \right]. \quad (6)$$

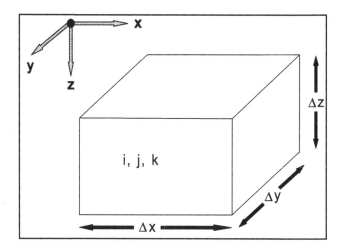

The bulk volume is

$$V_b = \Delta x \Delta y \Delta z \tag{7}$$

and may be substituted in Eq. (6).

The production term is defined as q and only exists for cells with wells. Injection is treated just like production with the exception that the sign is reversed.

The accumulation term represents the change in a cell with respect to time,

$$\text{accumulation} = \frac{\partial}{\partial t}\left(\frac{S\,V_b\,\phi}{5.615B}\right). \tag{8}$$

By substituting Eqs. (6) and (8) into Eq. (1), and using the definition of production, for the oil phase we can write,

$$0.00113\,V_b\left[\frac{\partial}{\partial x}\left(\frac{k\,k_{ro}}{\mu_o\,B_o}\frac{\partial\Phi_o}{\partial x}\right) + \frac{\partial}{\partial y}\left(\frac{k\,k_{ro}}{\mu_o\,B_o}\frac{\partial\Phi_o}{\partial y}\right)\right.$$
$$\left. + \frac{\partial}{\partial z}\left(\frac{k\,k_{ro}}{\mu_o\,B_o}\frac{\partial\Phi_o}{\partial z}\right)\right] - q_o = \frac{\partial}{\partial t}\left(\frac{S_o\,V_b\,\phi}{5.615\,B_o}\right). \tag{9}$$

The water phase equation is identical to Eq. (9) except all "*o*" subscripts become "*w*"s.

For the gas phase, accounting for the dissolved gas (in oil) and free gas,

flow in − flow out

$$
= 0.00113\, V_b \left[\frac{\partial}{\partial x} \left(\frac{k\, k_{rg}}{\mu_g\, B_g} \frac{\partial \Phi_g}{\partial x} \right) + \frac{\partial}{\partial y} \left(\frac{k\, k_{rg}}{\mu_g\, B_g} \frac{\partial \Phi_g}{\partial y} \right) \right.
$$

$$
+ \frac{\partial}{\partial z} \left(\frac{k\, k_{rg}}{\mu_g\, B_g} \frac{\partial \Phi_g}{\partial z} \right) + \frac{\partial}{\partial x} \left(\frac{k\, k_{ro}\, R_s}{\mu_o\, B_o} \frac{\partial \Phi_o}{\partial x} \right)
$$

$$
\left. + \frac{\partial}{\partial y} \left(\frac{k\, k_{ro}\, R_s}{\mu_o\, B_o} \frac{\partial \Phi_o}{\partial y} \right) + \frac{\partial}{\partial z} \left(\frac{k\, k_{ro}\, R_s}{\mu_o\, B_o} \frac{\partial \Phi_o}{\partial z} \right) \right]. \tag{10}
$$

Also, defining the total gas production,

$$
\text{gas production} = q_g + q_o\, R_s \tag{11}
$$

and the accumulation term as in Eq. (8), the total gas phase equation becomes

$$
0.00113\, V_b \left[\frac{\partial}{\partial x} \left(\frac{k\, k_{rg}}{\mu_g\, B_g} \frac{\partial \Phi_g}{\partial x} \right) + \frac{\partial}{\partial y} \left(\frac{k\, k_{rg}}{\mu_g\, B_g} \frac{\partial \Phi_g}{\partial y} \right) \right.
$$

$$
+ \frac{\partial}{\partial z} \left(\frac{k\, k_{rg}}{\mu_g\, B_g} \frac{\partial \Phi_g}{\partial z} \right) + \frac{\partial}{\partial x} \left(\frac{k\, k_{ro}\, R_s}{\mu_o\, B_o} \frac{\partial \Phi_o}{\partial x} \right)
$$

$$
\left. + \frac{\partial}{\partial y} \left(\frac{k\, k_{ro}\, R_s}{\mu_o\, B_o} \frac{\partial \Phi_o}{\partial y} \right) + \frac{\partial}{\partial z} \left(\frac{k\, k_{ro}\, R_s}{\mu_o\, B_o} \frac{\partial \Phi_o}{\partial z} \right) \right] - (q_g + q_o\, R_s)
$$

$$
= \frac{\partial}{\partial t} \left(\frac{S_g\, V_b\, \phi}{5.615\, B_g} + \frac{S_o\, R_s\, V_b\, \phi}{5.615\, B_o} \right). \tag{12}
$$

If we want to account for capillary pressure and gravity effects, the oil potential may be defined (for an incompressible fluid) as

$$\Phi_o = P_o - \frac{\rho_o D}{144}. \tag{13}$$

The water–oil capillary pressure is

$$P_{c_{wo}} = P_o - P_w \tag{14}$$

so the water potential (in terms of oil pressure) is

$$\Phi_w = P_o - P_{c_{wo}} - \frac{\rho_w D}{144}. \tag{15}$$

Similarly, the gas–oil capillary pressure is

$$P_{c_{go}} = P_g - P_o \tag{16}$$

resulting in the gas potential

$$\Phi_g = P_o + P_{c_{go}} - \frac{\rho_g D}{144}. \tag{17}$$

If we write Eq. (9) for the water phase considering only the x-direction and substituting Eq. (15) for the water potential, we have

$$0.00113\, V_b \left[\frac{\partial}{\partial x} \left(\frac{k\, k_{rw}}{\mu_w\, B_w} \frac{\partial \left(P_o - P_{c_{wo}} - \frac{\rho_w D}{144} \right)}{\partial x} \right) \right] - q_w$$

$$= \frac{\partial}{\partial t} \left(\frac{S_w\, V_b\, \phi}{5.615\, B_w} \right). \tag{18}$$

To simplify matters, let us define a mobility,

$$M_w = \frac{k\, k_{rw}}{\mu_w\, B_w} \tag{19}$$

and divide both sides of Eq. (18) by bulk volume (since $\partial V_b/\partial t = 0$), we obtain

$$0.00113 \left[\frac{\partial}{\partial x} \left(M_w \frac{\partial \left(P_o - P_{c_{wo}} - \frac{\rho_w D}{144} \right)}{\partial x} \right) \right] - \frac{q_w}{V_b}$$

$$= \frac{\partial}{\partial t} \left(\frac{S_w\, \phi}{5.615\, B_w} \right). \tag{20}$$

Now, let us work with the left side of Eq. (20), excluding the production term. We can define the partial differential by a finite-difference approximation

$$\frac{\partial A}{\partial x} \approx \frac{A_2 - A_1}{x_2 - x_1} \tag{21}$$

so that the left side of Eq. (20) becomes

$$\frac{0.00113}{\Delta x_i} \left[M_{w_{i+\frac{1}{2}}} \left(\frac{\partial \left(P_o - P_{c_{wo}} - \frac{\rho_w D}{144} \right)}{\partial x} \right)_{i+\frac{1}{2}} \right.$$

$$\left. - M_{w_{i-\frac{1}{2}}} \left(\frac{\partial \left(P_o - P_{c_{wo}} - \frac{\rho_w D}{144} \right)}{\partial x} \right)_{i-\frac{1}{2}} \right]. \tag{22}$$

If we look at the potential term using the partial of a sum, we can say that

$$\frac{\partial \left(P_o - P_{c_{wo}} - \frac{\rho_w D}{144} \right)}{\partial x} = \frac{\partial P_o}{\partial x} - \frac{\partial P_{c_{wo}}}{\partial x} - \frac{\partial \left(\frac{\rho_w D}{144} \right)}{\partial x} \tag{23}$$

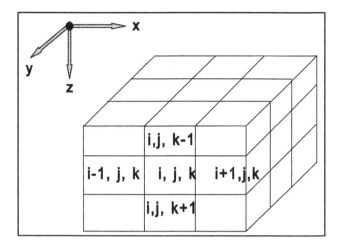

and the gravity term is the partial of a product

$$\frac{\partial\left(\frac{\rho_w D}{144}\right)}{\partial x} = \frac{\rho_w}{144}\frac{\partial D}{\partial x} + \frac{D}{144}\frac{\partial \rho_w}{\partial x} + \rho_w D\frac{\partial \frac{1}{144}}{\partial x}$$

$$= \frac{\rho_w}{144}\frac{\partial D}{\partial x} + \frac{D}{144}\frac{\partial \rho_w}{\partial x}.$$

$$(24)$$

A common simplifying assumption is that the gravity term is primarily a function of height and may be approximated (for an incompressible fluid) by

$$\partial\left(\frac{\rho_w D}{144}\right) \approx \frac{\rho_w}{144}\frac{\partial D}{\partial x}. \tag{25}$$

Using the finite difference approximation for $M_{i+1/2}$ (at the right interface of a cell),

$$\left(\frac{\partial P_o}{\partial x}\right)_{i+\frac{1}{2}} \approx \frac{P_{o_{i+1}} - P_{o_i}}{\frac{\Delta x_{i+1} + \Delta x_i}{2}}. \tag{26}$$

Similarly, for $M_{i-1/2}$ (at the left interface of a cell),

$$\left(\frac{\partial P_o}{\partial x}\right)_{i-\frac{1}{2}} \approx \frac{P_{o_i} - P_{o_{i-1}}}{\frac{\Delta x_i + \Delta x_{i-1}}{2}}. \tag{27}$$

When we substitute these terms into (22), we obtain

$$\frac{0.00113}{\Delta x_i} \left[M_{w_{i+\frac{1}{2}}} \frac{P_{o_{i+1}} - P_{o_i} - P_{c_{wo_{i+1}}} + P_{c_{wo_i}} - \frac{\rho_w}{144}(D_{i+1} - D_i)}{\frac{\Delta x_{i+1} + \Delta x_i}{2}} \right.$$

$$\left. - M_{w_{i-\frac{1}{2}}} \frac{P_{o_i} - P_{o_{i-1}} - P_{c_{wo_i}} + P_{c_{wo_{i-1}}} \frac{\rho_w}{144}(D_i - D_{i-1})}{\frac{\Delta x_i + \Delta x_{i-1}}{2}} \right]. \tag{28}$$

(28) can be expanded to include the y- and z-directions, and the oil and gas phase equations.

If we separate the right side of Eq. (20), we have

$$\frac{\partial}{\partial t} \frac{S_w \phi}{5.615 B_w}$$

$$= \frac{1}{5.615} \left(\frac{\phi}{B_w} \frac{\partial S_w}{\partial t} + S_w \phi \frac{\partial \frac{1}{B_w}}{\partial t} + \frac{S_w}{B_w} \frac{\partial \phi}{\partial t} \right), \tag{29}$$

and applying the chain rule

$$\frac{\partial A}{\partial t} = \frac{\partial A}{\partial P} \frac{\partial P}{\partial t} \tag{30}$$

to the last two terms of (29), we obtain

$$\frac{\phi}{5.615 B_w} \frac{\partial S_w}{\partial t} + \left(\frac{S_w \phi}{5.615} \frac{\partial \frac{1}{B_w}}{\partial P_o} + \frac{\frac{S_w}{B_w}}{5.615} \frac{\partial \phi}{\partial P_o} \right) \frac{\partial P_o}{\partial t}. \tag{31}$$

Employing the chain rule and the reciprocal derivative definition,

$$\frac{\partial \frac{1}{B_w}}{\partial P_o} = \frac{\partial \frac{1}{B_w}}{\partial B_w} \frac{\partial B_w}{\partial P_o} = -\frac{1}{B_w^2} \frac{\partial B_w}{\partial P_o}. \tag{32}$$

Using the definition of water compressibility

$$c_w = -\frac{1}{B_w} \frac{\partial B_w}{\partial P_o}, \tag{33}$$

and formation compressibility

$$c_f = \frac{1}{\phi} \frac{\partial \phi}{\partial P_o}, \tag{33a}$$

(31) can be rewritten as

$$\frac{\phi}{5.615\,B_w} \frac{\partial S_w}{\partial t} + \frac{S_w\,\phi\,c_w + S_w\,\phi\,c_f}{5.615\,B_w} \frac{\partial P_o}{\partial t}. \tag{34}$$

If we multiply the entire water phase equation by its formation volume factor, (34) simplifies to

$$\frac{\phi}{5.615} \frac{\partial S_w}{\partial t} + \frac{S_w\phi\,c_w + S_w\,\phi\,c_f}{5.615} \frac{\partial P_o}{\partial t}. \tag{35}$$

Since we are going to add the three phase equations together, we know that the sum of the first term of each phase in (35) will be

$$\frac{\phi}{5.615} \left(\frac{\partial S_w}{\partial t} + \frac{\partial S_o}{\partial t} + \frac{\partial S_g}{\partial t} \right) = \frac{\phi}{5.615} \frac{\partial \left(S_w + S_o + S_g \right)}{\partial t}. \tag{36}$$

However, the sum of the saturations is equal to one,

$$S_w + S_o + S_g = 1, \tag{37}$$

and the derivative of a constant is equal to zero,

$$\frac{\partial(S_w + S_o + S_g)}{\partial t} = \frac{\partial(1)}{\partial t} = 0, \tag{38}$$

so we can forget about the first term in (35) when solving for pressure in an IMPES formulation. Again, using the finite difference approximation,

$$\frac{\partial P_o}{\partial t} \approx \frac{P_o^{t+\Delta t} - P_o^t}{\Delta t}, \tag{39}$$

so that putting it all together for the water phase in the x-direction, assuming we add the three phase equations together and remembering that we multiplied by B_w in (35), we get (using (28), (35), (39) and the production term as defined in (9), $-q_w B_w / V_b$):

$$\frac{0.00113\, B_{w_i}}{\Delta x_i} \left[M_{w_{i+\frac{1}{2}}} \frac{P_{o_{i+1}} - P_{o_i} - P_{c_{wo_{i+1}}} + P_{c_{wo_i}} - \frac{\rho_w}{144}(D_{i+1} - D_i)}{\frac{\Delta x_{i+1} + \Delta x_i}{2}} \right.$$
$$\left. - M_{w_{i-\frac{1}{2}}} \frac{P_{o_i} - P_{o_{i-1}} - P_{c_{wo_i}} + P_{c_{wo_{i-1}}} - \frac{\rho_w}{144}(D_i - D_{i-1})}{\frac{\Delta x_i + \Delta x_{i-1}}{2}} \right] - \frac{q_w B_w}{V_b}$$
$$= \frac{S_w \phi c_w + S_w \phi c_f}{5.615} \frac{P_o^{t+\Delta t} - P_o^t}{\Delta t}. \tag{40}$$

This results in one equation for each cell in the model and in adding the phase equations together (and eliminating the saturation derivatives with respect to time), we have constructed the **IMPES** formulation.

If we maintain separate phase equations, we have a **fully implicit** formulation with three equations per cell; in this formulation, the saturation derivatives with respect to time cannot be eliminated and must be approximated using a finite difference.

The previous derivation, with several options, is shown in both Chapter 6 and Appendices B and C of M-13.

Nomenclature

- k absolute permeability, md
- k_r relative permeability
- μ viscosity, cp
- ρ density, lb/ft^3
- P pressure, psi
- Φ potential, psi
- D depth, ft
- x length, ft
- Δx cell length, ft
- y width, ft
- Δy cell width, ft
- z thickness, ft
- Δz cell thickness, ft
- Δt timestep, days
- S saturation, fraction
- ϕ porosity, fraction
- c compressibility, 1/psi
- B formation volume factor, RVB/STB for liquids, RVB/MCF for gas
- Q, q rate (+ for production, − for injection), STB/day for liquids, MCF/day for gas
- V volume, ft^3
- A area, ft^2

     Subscripts

- b bulk
- x, y, z directional notation
- o oil
- g gas
- w water
- f formation
- i, j, k indices of directional location

Superscripts

- t old (current) time level
- $t + \Delta t$ new time level

2.2. Types of Simulators

In developing the flow equations as shown in Eq. 40, the three phases were added together to obtain one final equation which has the general form:

$$\text{LHS Coeff.} * \text{Pressures} + \text{Constants} = \text{RHS Coeff.} * \text{Pressures}$$

$$\frac{0.00113\, B_{w_i}}{\Delta x_i} \left[M_{w_{i+\frac{1}{2}}} \frac{P_{o_{i+1}} - P_{o_i} - P_{c_{wo_{i+1}}} + P_{c_{wo_i}} - \frac{\rho_w}{144}\left(D_{i+1} - D_i\right)}{\frac{\Delta x_{i+1} + \Delta x_i}{2}} \right.$$

$$\left. - M_{w_{i-\frac{1}{2}}} \frac{P_{o_i} - P_{o_{i-1}} - P_{c_{wo_i}} + P_{c_{wo_{i-1}}} - \frac{\rho_w}{144}\left(D_i - D_{i-1}\right)}{\frac{\Delta x_i + \Delta x_{i-1}}{2}} \right] - \frac{q_w B_w}{V_b}$$

$$= \frac{S_w\, \phi\, c_w + S_w\, \phi\, c_f}{5.615} \frac{P_o^{t+\Delta t} - P_o^{t}}{\Delta t},$$

where the pressures associated with the left-hand side of the equation are locationally-dependent and those on the right-hand side are time-dependent; however the pressures on the left-hand side require that some time level be assigned to them. If we express this equation as

$$\text{LHS Coeff.} * \text{Pressures}^{\text{old}} + \text{Constants}$$

$$= \text{RHS Coeff.} * \left(\text{Pressures}^{\text{new}} - \text{Pressures}^{\text{old}}\right)$$

the formulation is known as an **explicit pressure** solution. This assumes that the coefficients on both sides do not change and that the old time-level pressures may be used on the left-hand side; this is obviously a very simple solution since it is a series of one-equation, one-unknown solutions. Unfortunately, since the coefficients on both sides involve pressure-dependent quantities and the old pressures may not reflect the same locational relationship on the left-hand side, this type of solution required very small time changes (known as timesteps). This fully explicit formulation is not useful due primarily to timestep limitations.

If we express the solution equation as

$$\text{LHS Coeff.} * \text{Pressures}^{\text{new}} + \text{Constants}$$
$$= \text{RHS Coeff.} * (\text{Pressures}^{\text{new}} - \text{Pressures}^{\text{old}})$$

we have an equation that accurately depicts the pressure changes but it becomes apparent that the coefficients should also be at the new time level or an inequality exists. So we can rewrite the equation as

$$(\text{LHS Coeff.} * \text{Pressures})^{\text{new}} + \text{Constants}$$
$$= (\text{RHS Coeff.} * \text{Pressures})^{\text{new}} - (\text{RHS Coeff.} * \text{Pressures})^{\text{old}},$$

where we don't know any of the variables involved except for the values at the *old* time level. Initially, this looks like a real problem, especially when one considers that the LHS includes a mobility term which has pressure-dependent variables (solution gas, formation volume factors and viscosities) and saturation-dependent variables (relative permeability and capillary pressures), and the RHS includes saturations as well as pressures. Basically, this means that if we can guess the correct pressures and saturations to determine the correct coefficients, we will have an equation (actually "n" equations in "n" unknowns) to solve. If we rearrange our equation to collect all of the *old* time level items on the RHS (they become constants) and move all of the *new* (unknown) items to the LHS, we have

$$(\text{LHS Coeff.} * \text{Pressures})^{\text{new}} - (\text{RHS Coeff.} * \text{Pressures})^{\text{new}}$$
$$= -\text{Constants} - (\text{RHS Coeff.} * \text{Pressures})^{\text{old}}.$$

In simplified form, the system of equations for "n" cells is

$$A\ P_0 + B\ P_1 + C\ P_2 \qquad = D$$
$$E\ P_1 + F\ P_2 + G\ P_3 \qquad = H$$
$$I\ P_2 + J\ P_3 + K\ P_4 \qquad = L$$
$$\ldots$$
$$S\ P_{n-2} + T\ P_{n-1} + U\ P_n \ = V$$
$$W\ P_{n-1} + X\ P_n + Y\ P_{n+1} = Z$$

and since the reservoir is a closed system $A = Y = 0$ and the system of "n" equations has "n" unknowns:

$$B\ P_1 + C\ P_2 \qquad\qquad = D$$
$$E\ P_1 + F\ P_2 + G\ P_3 \qquad = H$$
$$I\ P_2 + J\ P_3 + K\ P_4 \qquad = L$$
$$\ldots$$
$$S\ P_{n-2} + T\ P_{n-1} + U\ P_n \ = V$$
$$W\ P_{n-1} + X\ P_n \qquad\quad = Y$$

For example, in a 5 cell model,

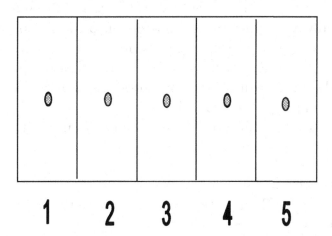

Five cell linear model

$$B \; P_1 + C \; P_2 \qquad = D$$
$$E \; P_1 \; + F \; P_2 + G \; P_3 = H$$
$$I \; P_2 \; + J \; P_3 + K \; P_4 = L$$
$$M \; P_3 \; + N \; P_4 + Q \; P_5 = R$$
$$S \; P_4 \; + T \; P_5 \qquad = V,$$

we have 5 equations with 5 unknowns $(P_1 \to P_5)$, which is certainly solvable (not much fun, but solvable).

If we assume that the saturations change only slightly during a timestep, then they (and their dependent quantities of relative permeability and capillary pressure) may be held constant during the pressure solution and calculated using a material balance once the correct pressures are known. This technique is known as the **IMPES (IMplicit Pressure, Explicit Saturation)** method and is the most common form used in field (cartesian) simulators (see Sec. C.5, p. 143, M-13). Since we have fixed the saturation-related variables, all we need to do is estimate the pressure-dependent variables and we get the right answer; unfortunately, getting all of these estimates correct takes several guesses, commonly known as **iterations**. If we start to guess correctly, both sides of the equation will approach each other and we will converge on the right answer (or very close to it). If we guess badly, the whole thing comes unglued (diverges) and the model blows up. This usually means that either we are guessing poorly or some data is in error; if we are guessing poorly, to reduce the guess required (the new pressure), we can reduce the timestep and hopefully, guess better.

A simple schematic of this procedure in an IMPES model would be

1. Guess Pressures.
2. Calculate PVT properties (for Flow Coefficients).
3. Calculate New Pressures.
4. Do New Pressures = Guessed Pressures?

No — Go back to Step 1 but SKIP Step 2 (Inner Iteration).
No — Go back to Step 1 (Outer Iteration).
Yes — Go on.

5. Calculate Saturations.
6. Go to the next timestep and start Step 1 again.

Variations in an IMPES model may be used to update saturations during the pressure solution; in this case the model is known as **semi-implicit**. A simple example would be to project new relative permeability values based on previous saturation changes; however, if the saturation profile is not consistent (as when shutting-in a well), the technique leads to errors. Note that a semi-implicit model still has "n" equations in "n" unknowns and the three phases are added together (see Sec. C.8, p. 146, M-13). Put very simply, a semi-implicit model is a GT version of the plain IMPES.

One popular type of variation is the **sequential** modification which combines the IMPES approach with a Buckley–Leverett saturation solution (see C.6, p. 144, M-13). A semi-implicit approach of the sequential solution estimates changes using various chord-slope relationships (again, see C.8). Technically, the sequential approach solves the pressure equation from an IMPES formulation and then uses those pressures to solve for saturations from single phase equations; this procedure is used in every iteration (or after a specified number of pressure iterations). While this technique allows larger timesteps, it may also yield unrealistic results and from an applied simulation viewpoint, unless one is well-versed in its use, the disadvantages tend to outweigh the advantages.

While the IMPES formulation works quite well for field studies, when large saturation changes occur over a timestep (as in a small cell), extremely tiny timesteps may be required. A better idea would be to solve implicitly for both pressures and saturations; this is the **fully implicit** method. Unfortunately, if we use the IMPES formulation, it results in "n" equations with "$3n$" unknowns (pressure and two of the three saturations). This one is too tough to solve; however, if we refrain from adding the oil, gas, and water equations together, we will have "$3n$" equations with "$3n$" unknowns which can be solved but requires a great deal of additional computer time (see C.7, p. 145, M-13).

A simple schematic of the solution procedure employed by a fully implicit model would be

1. Guess Pressures and Saturations.
2. Calculate PVT properties, P_c, and k_{rel} (for Flow Coefficients).
3. Calculate New Pressures and Saturations.
4. Do New Pressures = Guessed Pressures
   and New Saturations = Guessed Saturations?

   No — Go back to Step 1 but SKIP Step 2 (Inner Iteration).
   No — Go back to Step 1 (Outer Iteration).
   Yes — Go on.

5. Go to the next timestep and start Step 1 again.

The most common use of a fully implicit model is in radial simulators where the cells near the wellbore are extremely small and may undergo large saturation changes (for example, when coning occurs). Another problem which may require a fully implicit model is that of rapid vertical gas movement (percolation). An attempt to use various degrees of an implicit solution (i.e., pressures and rapidly changing saturations) in only those cells undergoing rapid changes is the **adaptive implicit method (AIM)**.

2.3. Solution Techniques

In solving the simulation equations, several different mathematical techniques are available. Although these methods technically consist of two types of matrix solutions (direct and iterative), all are iterative in use due to the fact that the coefficients must be correct in order to have an equality. A **direct solution** will solve a system of equations in one set of calculations while an **iterative solution** will use some sort of predictor–corrector technique to converge on the answer. The primary direct solution employed in reservoir simulators is gaussian elimination (GAUS) while the most popular iterative techniques are the strongly implicit procedure (SIP), successive over-relaxation methods (SOR), and conjugate gradient-like methods (CGL). One additional iterative technique that is still mentioned, but should seldom be used is the alternating-direction implicit procedure (ADI).

When using iterative solution techniques, in order to converge rapidly (or converge at all), various weighting factors are used during the iterations: these factors are known as iteration parameters and weight-average the new guess of pressures. If the solution is converging on the correct pressure, the previously iterated pressure should be heavily weighted (called an **inner iteration** or a **linear iteration**, or sometimes, simply an **iteration**); however, if the solution is not converging, the pressure from the previous timestep should be heavily weighted. Although all recalculations are iterations, a common plan of attack is to weight previously iterated pressures (assuming the solution is converging) during inner iterations, but after a certain number of these iterations, to re-evaluate the coefficients in order to obtain a more accurate system of equations to solve (usually considered an **outer iteration** or a **nonlinear iteration** or a **cycle**). So in essence, we attempt to converge using each previous guess and after awhile, if we don't get close enough, we recalculate the flow coefficients.

Each inner iteration is an attempt to solve a system of linear equations, so when using an iterative method, several inner iterations would be required; however, when using a direct solution method, only one inner iteration would be needed. Each outer iteration is an attempt to solve the actual nonlinear equation by re-evaluating the flow coefficients. The equation is solved when the LHS of the equation is equal to the RHS, or in actual practice, when they are close enough. This minimal error (when they are close enough) is called the **residual** (This author prefers "residue": what's left over) and is defined as

$$|\text{LHS} - \text{RHS}|.$$

For a simulation grid involving several wells, we usually define the residual as being normalized by the flow rates of the wells

$$\frac{|\text{LHS} - \text{RHS}|}{|\text{Flow Rates}|}.$$

A solution yielding a value of zero for the residual would indicate an exact solution; however, due to the changing of coefficients and pressures, we will not obtain an exact zero very often; also, the extra number of iterations required to get to zero is not justified from a computer time standpoint. We define some tolerance as close enough to zero to be acceptable; obviously, the smaller the tolerance, the better the pressures (but more iterations are required). Even when using a direct solution, some residual will usually exist. An acceptable tolerance for many black oil simulations is 0.005. A very minimal discussion of residuals is on p. 148, M-13.

There are other parameters that also indicate the "goodness" of the solution. One of these values is the **change in pressures between iterations**; while this value does not indicate good answers, it does indicate that the answers fit the equation or are converging very slowly. For an IMPES model, a value of less than 0.1 psi is preferred (see p. 148, M-13).

The **maximum changes in saturations and pressures during a timestep** are other solution indicators. In general, when these values are large, the model tends to "average-out" or slough-over distinct occurrences, and becomes merely a glorified material balance. It is possible to take a ten-year timestep; however, the results would seldom match the same simulation using one month timesteps. When saturation and pressure changes over a timestep exceed reasonable values, the timestep size should be reduced. If these changes are larger, linear interpolations for relative permeabilities and PVT values will be in error, and we may not be solving a representative equation. Reasonable values for maximum saturation changes are 4.0% and for pressure, 150 psi. These changes should normally occur in well cells. Some models also employs target values of pressure and saturation change to adjust timestep size to achieve the desired target values.

Other indicators are **material balances** calculated for the oil, gas and water phases. Two types of material balances may be employed: cumulative and incremental. The cumulative material balance is usually defined as

$$\frac{|\text{Original-in-place} - \text{Current-in-place}|}{|\text{Net Cumulative Production}|}$$

and is an overall indicator of the accuracy of the run. It is not sensitive to individual timesteps in that a poor solution for one timestep will be overshadowed by several "good" timesteps. In the equation shown, a value as close to 1.0 is desired, with accuracy of 0.001 preferred. In most simulators, for 3-phase flow, the oil and water saturations are calculated and the gas saturation is obtained by difference; as such, a greater error is allowable in the gas balance. There are two occasions when the cumulative material balance would be expected to be meaningless: 1. early in the simulation when the differences between original and current-in-place numbers vary slightly and the cumulative production is small and 2. at any point during the run when large volumes of fluid have been injected (since the net cumulative production is cumulative production less cumulative injection). In general, the first condition applies until 0.5% of the reservoir is produced.

The **incremental material balance** is indicative of the condition during a timestep and is defined as

$$\frac{|\text{Volume at start of timestep} - \text{Volume at end of timestep}|}{|\text{Net production during timestep}|}.$$

Again, this would be a number near 1.0, but quite frequently to distinguish this value from cumulative values, it is subtracted from 1.0 yielding a number approaching zero. Large variations in this value indicate either timestep problems or saturation pressure changes.

Returning to the solution methods, we need to understand the various options that exist. From an applied simulation viewpoint, we want a method that is reasonably fast and accurate, and always converges to the right answer without requiring intervention on our part; the term that mathematicians use for this criteria is **robust** (I suppose if it's really good, it also has gusto). As mentioned previously, most iterative techniques employ iteration parameters (weighting factors) to speed up the rate of convergence.

The **Alternating Direction Implicit procedure (ADI)** was the first popular solution technique used in reservoir simulators. It essentially divided the timestep into "half timesteps" (for 2-D models) and solved in the x-direction for the first half of the timestep; it then solved in the y-direction for the second half of the timestep using pressures obtained as a result of the x-sweep. ADI was satisfactory in that it did not require much computer storage; however, in order to obtain satisfactory results, various weighting factors are used during the iterations and the selection of these parameters was critical. From application point of view, this was not very popular. Suffice to say, that the ADI technique has been replaced by SIP, SOR or CGL methods in all good models (see Sec. 6.7.3, p. 69, M-13).

The **Strongly Implicit Procedure (SIP)** is an iterative approximative factorization technique which does not require a large amount of storage, converges rapidly (usually) and is fairly insensitive to iteration parameters. It alters the original equations to yield a matrix which may easily be factored into an upper and lower set of matrices. Being somewhat insensitive to iteration parameters, it does not require an extensive mathematical background to use and is recommended for typical simulation studies, particularly studies involving more than 500 cells. Recommended iteration parameters for SIP vary from 0.0 to 1.0. In general, when the pressure solution "blows up" using SIP, changing the iteration parameters will not help. The major complaint against SIP is that if the solution tolerance is not small enough, asymmetrical results will occur in a symmetrical system. For black oil studies, this problem appears minimal (except at extremely unfavorable mobility ratios) when the previously recommended tolerance of 0.005 is used, although a lower value may be required for 4-component models. When in doubt, rerun a given simulation using a lower tolerance to see if the answers change significantly; this is a form of sensitivity testing. For additional discussion of SIP, see Secs. 6.7.4.1 (p. 69) and C.9.2.3 (p. 149) of M-13.

Successive Over-Relaxation (SOR) techniques are another type of iterative method. They maintain symmetry and solve (or fail to solve) the pressure equation about as often as the SIP technique. The weighting factor in these techniques is called a successive relaxation parameter, Ω (omega), and selection of the proper value of Ω will result in a more rapid convergence with SOR than with SIP; however, failure to select an optimum Ω value may result in a more time consuming solution using SOR. Also, the optimum value of Ω may change during a simulation run. Some of these problems may be overcome by various algorithms for determining a proper value of the relaxation parameter. Ω is usually a number having a value between 1.0 and 2.0

for over-relaxation. The basic SOR prediction equation is

$$P_{new} = P_{old} + \Omega(P_{calc.} - P_{old}).$$

The various types of relaxation methods are point successive overrelaxation (PSOR), line successive overrelaxation (LSOR), and block successive overrelaxation (BSOR); a slightly different version is slice successive overrelaxation (SSOR). The first word in each type denotes the number of cells which are relaxed at one time. M-13 explains SOR methods in Secs. 6.7.2 (p. 68), C.9.2.1 and C.9.2.2 (p. 148).

Conjugate Gradient-Like (CGL) methods are minimization type of iterative solution methods employed in reservoir simulators. These methods use a preconditioning matrix to approximate the inverse matrix and in solving the equations, various acceleration schemes may be employed for rapid convergence. A more recent improvement for preconditioning the matrix is the use of nested factorization or other incomplete lower-upper triangularization (ILU) matrices. CGL methods may be used in conjunction with other solution techniques and are discussed in Secs. 6.7.4.2, 6.7.4.3 (p. 71) and C.9.2.4 (p. 150) of M-13.

Gaussian Elimination (GAUS) is a direct solution method (thus employing only one inner iteration), but it requires a certain number of outer iterations so that the coefficients define the correct equation. For all problems involving reasonable data, this method will solve in fewer total iterations than SIP, SOR or CGL methods; however, the amount of computer storage and the number of calculations per iteration are greater for gaussian elimination than for the other techniques. Thus one iteration in GAUS may require more computer time than several iterations in an iterative method. A large amount of computer storage can be saved by use of reordering schemes, the most common of which are known as the **alternate diagonal (D4)** and

checkerboard (A3) ordering schemes as shown in Fig. 6.5, p. 67, M-13. For studies involving fewer than 500 cells, gaussian elimination should be employed; for studies in the range of 500–2000 cells, GAUS should be considered as feasible depending on the particular computer system; again, it is probably desirable to sensitivity test GAUS against an iterative method. Since gaussian elimination solves directly and should converge very rapidly, and since iterative methods are fast when they have good estimations, a combination of GAUS and SIP yields good initial convergence and rapid final convergence. The sequence of using GAUS to first generate good estimations of pressures and then follow up by SIP yields a good solution scheme; many times, satisfactory solutions are obtained without ever requiring the SIP solution. Also, if the direct solution does not converge within a small number of iterations, usually we have a pretty tough equation to solve and repeated direct methods will be no better than iterative ones. In general, from 3 to 5 iterations using gaussian elimination should be sufficient. Additional information on gaussian elimination may be found in Secs. 6.7.1 (p. 64) and C.9.1 (p. 146) of M-13.

An outline for **selecting a solution technique** is given on p. 72 of M-13. Often, the choices available are not as complete as those listed in the monograph. Simply put,

1. for a minimum number of cells (usually <500, but dependent on the computer system) and a hard problem, use GAUS;
2. for a minimum number of cells and an easy problem, use GAUS or any iterative technique (whichever is faster);
3. for a large number of cells, use an iterative method (usually whatever is available)

    a. for cross-sections, SOR is usually faster than SIP;
    b. for 3-D, SIP is usually faster than SOR;
    c. for areal, it's a toss up;
    d. **if you have a CGL method, it's probably better than SIP or SOR.**

Problems — Chapter 2

1. *Derivation of simulation flow equations.* Using Eq. (9) which is for oil phase flow, develop a final finite difference form for use in a simulator for flow in the y-direction. Use Eq. (40) as a guide.

2. *Iterative & direct solution techniques.* Solve the following set of equations using both iterative and direct solution techniques.

$$\text{I.} \quad 2x + y = 4$$
$$\text{II.} \quad x + y + z = 6$$
$$\text{III.} \quad 3y - z = 3$$

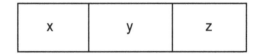

Use the following procedure for the iterative solution:

1. Guess z (use 0 for the first guess).
2. Calculate y using Equation III.
3. Calculate x using Equation I with y from the previous step.
4. Calculate the error (residual) using Equation II where error = RHS − LHS.
5. If error ≈ 0, you're done; if not, calculate a new guess for z with Equation II and return to step 2.

NOTE: Do not use more than 7 iterations.

Use the following procedure for the direct solution:

1. Solve Equation I for y.
2. Substitute this result in Equation II.
3. Solve Equation III for y.
4. Substitute this result in Equation I.
5. Solve the resultant equation from step 4 for z.
6. Substitute this result in the equation from step 2 and solve for x.

7. Solve Equation I for y.
8. Solve Equation III (or Equation II) for z.

3. *Iterative solution techniques weighting factors.* Another reasonable choice to solve the previous set of equations is to guess the value of y.

$$\text{I.} \quad 2x + y = 4$$
$$\text{II.} \quad x + y + z = 6$$
$$\text{III.} \quad 3y - z = 3$$

Procedure:

1. Guess y (use 0 as a first guess).
2. Calculate x from Equation I.
3. Calculate z from Equation III.
4. Calculated the error using Equation II where error = RHS − LHS.
5. If error $\approx$0, you're done; if not, calculate a new guess for y with Equation II and return to step 2.

NOTE: Do not use more than 4 iterations.

Chapter 3

PVT DATA

Although the main PVT (Pressure, Volume, Temperature) requirements were explained under the section on basic data requirements a few of the concepts and simulator requirements will bear some additional consideration.

Repressurization involves an increase in pressure in any portion of the reservoir. In simulation, it is concerned with an increase in pressure in any cell. Obviously, injection well cells will undergo repressurization, but note that shutting in a producer will also cause a repressurization effect in the producing well cell. Two factors will influence the result of this effect: the amount of the pressure increase and the amount of free gas available to be dissolved. Three PVT properties are affected by repressurization: oil formation volume factor (B_o), solution gas (R_s), and oil viscosity (μ_o).

Under normal pressure depletion, solution gas remains constant above the bubble point and drops as pressure drops below bubble point. If the pressure increases, the amount of dissolved gas increases as long as free gas is available to be dissolved in the oil. Once the original amounts of oil and gas in the reservoir are altered by production, the original solution gas data may not apply. The bubble point pressure may be defined in either of two ways: 1. for an undersaturated reservoir, it is the pressure at which the first bubble of gas comes out of solution; 2. for a saturated reservoir, it is the pressure at the gas–oil contact. The

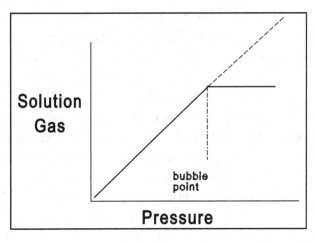

Solution gas plot

use of the term "bubble point pressure" is preferred in describing initial reservoir conditions; once these conditions are altered, the pressure at which all free gas could be dissolved may be referred to as **saturation pressure** (which is essentially a dynamic bubble point pressure).

If we produce a portion of an initially undersaturated reservoir (point A) below the bubble point (point B) and then

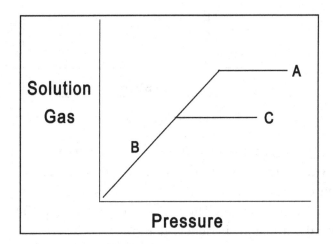

Water injection example

increase the pressure by water injection (point C), some of the released solution gas will probably be produced with the oil and the saturation pressure will be lower than the original bubble point pressure. Implicit in this example is that additional gas is produced (as in a completion at the top of a reservoir).

On the other hand, if we produce a portion of an undersaturated reservoir (A) below the bubble point (B) and then increase the pressure by injecting gas of similar composition to the reservoir gas point (C), the solution gas curve continually increases due to the fact that there is free gas available to be dissolved at the higher pressures. Note that the line B-C indicates a saturated condition: all of the gas that can be dissolved at a given pressure is in solution; if additional gas is present, it will exist as free gas.

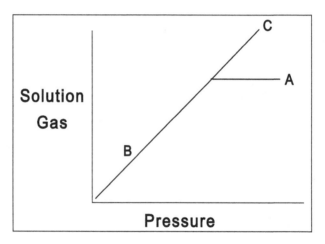

Gas injection example

This continually increasing solution gas curve is the **repressurization data**. For similar compositions, it should be an extension of the original data. If the injected gas is leaner than the original reservoir gas, the repressurization curve falls below the original extension; if richer, it moves above.

Now let us look at a more complicated example involving repressurization. A portion of an undersaturated reservoir (A)

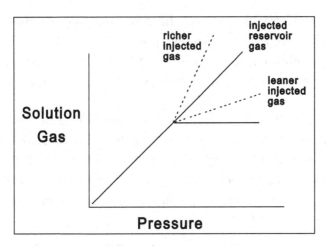

Effect of gas composition
Solution gas plot

is produced below the bubble point (B); the wells are completed low on structure, so more oil than freed gas is produced. Water injection is begun and the pressure is increased to discovery pressure (C); additional water injection dissolves all of the remaining free gas and results in an undersaturated condition (D). Gas injection is now initiated; note that the pressure will not increase until the amount of gas that could be dissolved at point E is reached; further injection with increasing pressure would result

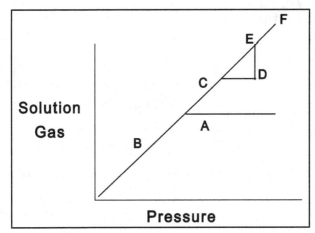

Complicated repressurization example
Solution gas plot

in point F. If we blow down the reservoir, the original path is not retraced, but simply follows points F-E-C-B.

The model employs this concept in every cell at each timestep by comparing at the dissolved gas, free gas, cell pressure and saturation pressure.

The concept is very similar for the oil formation volume factor. Whereas the solution gas curve becomes horizontal when an undersaturated condition exists, the oil formation volume factor must "break back" based on the undersaturated oil compressibility. Generalizing an equation for the undersaturated oil formation volume factor we have

$$B_o = B_{oSP}\, e^{-c_o(P - P_{SP})} \approx B_{oSP}[1 - c_o(P - P_{SP})],$$

where SP indicates saturation pressure. To continue our complicated example on the oil formation volume factor curve, results in the figure shown below.

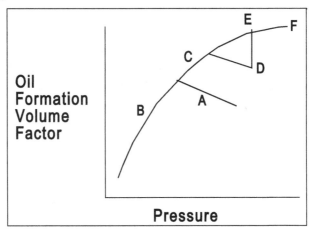

Complicated repressurization example
Oil formation volume factor plot

The discussion concerning injecting leaner or richer gas also applies.

Note that projecting the oil formation volume factor curve is not as simple as projecting the solution gas plot due to the curvature involved. One very common problem that occurs when using extrapolated data is that $B_o < R_s\, B_g$ (a physical impossibility). Obviously, laboratory measured data is preferred but is not always available. If you are studying injection processes, you should plan on obtaining repressurization data at injection pressures from your PVT analysis.

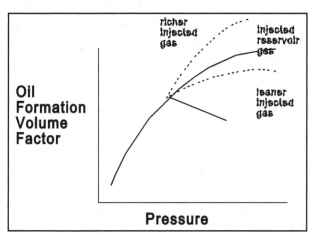

Effect of gas composition
Oil formation volume factor plot

At this point the concept of oil viscosity repressurization is fairly straightforward; again using our more complicated

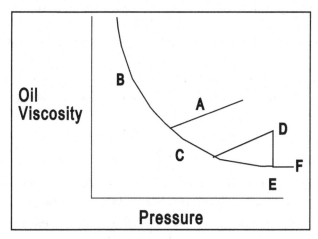

Complicated repressurization example
Oil viscosity plot

example we can illustrate this effect. The repressurization data in this case is on the curve BCEF.

Equal-spacing of PVT data is a technique used in models to speed up the determination of fluid properties which are required for every cell and iteration. Without equal-spacing, the model must search through the pressure entries until it locates the two values bracketing the cell pressure. With equally-spaced PVT entries, the point at which to enter the PVT table is directly calculated from

$$1 + \frac{(P - P_{1 \text{ table}})}{\text{spacing}}$$

and the decimal portion indicates the relative position of the actual pressure with respect to the tabular values. Since many models operate on this basis, that leaves you with two options: enter your data already equally-spaced or let the model create the equal-spacing. In the former case, your data is unaltered and all of your pressure entries are retained. Note that one of your pressure values must be the bubble point pressure. To illustrate the number of table searches required, for a 1000 cell model averaging 3 non-linear iterations per timestep, employing 20 timesteps per year for 10 years of simulation, we have

$$1000 * 3 * 20 * 10 = 600{,}000 \text{ PVT table searches.}$$

For data that is not equally-spaced, knowing that the bubble point pressure is a required entry, the equal-spacing increment is

$$\text{INC} = \frac{(P_{BP} - P_{1 \text{ table}})}{(N_{BP} - 1)}$$

where N_{BP} is the number of the bubble point entry. Allowing the simulator to equally-space PVT data can lead to problems in two different ways: either some of the higher pressures are not retained or pressures greater than the data entered are needed.

For example, the following input pressures:

> 100
> 1400
> 2100 - bubble point
> 5000
> 6000

would result in equally-spaced entries of

> 100
> 1100
> 2100
> 3100
> 4100.

The simple solution in this case is to add dummy (but realistic) values of 6050 and 6100 to the original data.

The second case would exist when input values of

> 100
> 1400
> 2100 - bubble point
> 2300
> 2800

were spaced to

> 100
> 1100
> 2100
> 3100
> 4100

and the solution here is to either add higher pressure entries or, most likely, to retain the shape of the curve, add values below the bubble point to reduce the spacing. Some models would terminate this table at 2100 psi; some will extrapolate and add additional entries.

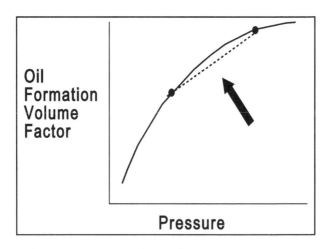

Linear interpolation of PVT data

Another more important problem that may exist due to equal-spacing is actual data changing due to the use of a straight line interpolation between data points. The only simple solution to this problem is to equally-space your PVT data (which results in the same data being generated after equal-spacing); otherwise, you must always compare the equally-spaced data to the original data, and add more definitive points where necessary. Automatic equal-spacing has probably caused as much grief as it has saved people the time required to manually equal-space their PVT entries.

Smoothness of PVT data is a necessity to avoid large amounts of pressure iterations and/or timestep reductions. Laboratory PVT data should be plotted and smoothed if necessary and automatically equally-spaced data should also be plotted and checked to determine that smoothness has been maintained. Extremely critical are values required when going through a saturation pressure. For example, remembering the plots of flash and differential oil formation volume factors, it is readily apparent that use of differential data will result in fewer iterations required for the pressure solution when passing through the bubble point.

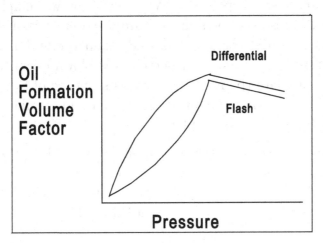

Flash and differential liberation

Remember, PVT entries should span the entire range of pressures expected during the simulation run.

Multiple sets of PVT tables may be used to model varying fluid properties as in either compositional effects or multiple reservoirs within a field. If compositional effects are pronounced, a compositional model should be used in lieu of a black oil simulator. PVT tables are assigned to groups of cells, and fluids traveling from one assigned group to another are instantly transformed to the properties of the new table (see Secs. 3.4.3 (p. 26) and 4.6.3 (p. 40) in M-13).

An oil (or gas) tracking feature allows monitoring the flow of fluids from one region to another and includes mixing of various fluids; tracer tracking, on the other hand, allows fluids to be tagged, but usually does not allow mixing.

Problems — Chapter 3

1. *Repressurization concepts.* A reservoir was discovered in an undersaturated condition (call this state, point A). Gas was injected to maintain pressure (point B), and eventually formed a secondary gas cap (point C). Water injection was employed to increase the reservoir pressure (above discovery pressure) until an undersaturated condition existed again (point D). Finally, all injection ceased and the reservoir was blown down to abandonment pressure (point E). Show this sequence of events on a solution gas (as a function of pressure) plot and an oil formation volume factor vs. pressure plot.

2. *Equally-spaced PVT tables.* In order to simulate a reservoir with a fair amount of pressure maintenance as a result of gas cycling (due to a limited gas market), the following pressures have been selected for use in the model PVT table:

$$400$$
$$1000$$
$$2000$$
$$3000$$
$$4000 \text{ - bubble point}$$
$$4100$$
$$4200$$
$$4300$$
$$4500$$
$$5000$$

Pressures are estimated to remain in the 4000 to 4500 psia range. Is the table shown appropriate for use when the data is equally-spaced?

If we elect to inject the produced gas into a different reservoir and follow normal pressure depletion in this reservoir, is our original pressure table adequate?

Chapter 4

RELATIVE PERMEABILITY AND CAPILLARY PRESSURE DATA

Previously, the primary concepts of relative permeability and capillary pressure were introduced; however, several additional aspects of these relationships are used extensively in simulation.

Hysteresis effects in both relative permeability and capillary pressure data occur when a flow reversal exists. To understand this concept, let us return first to the basic idea behind wettability which is that it is the degree to which a fluid wets a solid surface. An increase in the wetting-phase is known as an **imbibition** process whereas a decrease in the wetting phase is a **drainage** process. Waterflooding a water-wet reservoir is

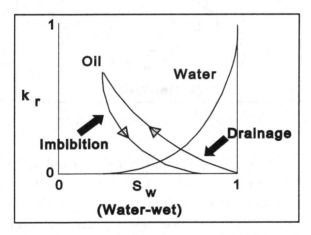

Relative permeability hysteresis

an imbibition process; back-producing an injector in this reservoir would result in a drainage process. Waterflooding an oil-wet system is a drainage process. Depending on the type of process that the reservoir is undergoing, the capillary pressures and non-wetting phase relative permeabilities may undergo a transformation. Although a slight variation may occur in the wetting phase relative permeability, this effect is usually considered to be negligible.

On a capillary pressure plot, the hysteresis effects (or the change between imbibition and drainage) for this system would appear as illustrated. The drainage curve shown in the relative permeability and capillary pressure graphs is a **primary drainage** curve and exists only when drainage occurs first. In a water-wet environment, you should equilibrate (initialize) using drainage data and then run the simulator using imbibition data.

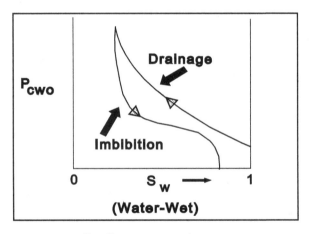

Capillary pressure hysteresis

When a drainage mechanism occurs after imbibition, a **secondary drainage** curve exists as shown for capillary pressure data. The secondary drainage curve will be in effect for all flow reversals initially occurring after primary imbibition (i.e., assuming a water-wet system).

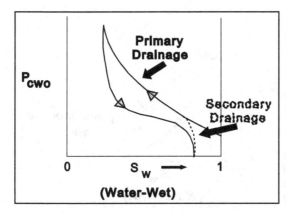

Capillary pressure secondary drainage

This secondary drainage curve for the relative permeability data creates an envelope with the imbibition curve as shown.

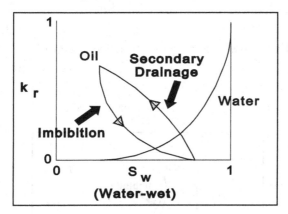

Relative permeability secondary drainage

However, these curves only exist when the saturation value is at either end point of the flow reversal. At intermediate saturations, the curve deviates at the point of reversal and follows an intermediate path which will eventually reach the correct end-point. What this means is that the model checks for a flow reversal based on a decrease in the wetting phase saturation over a timestep and adjusts the relative permeability accordingly. A similar adjustment is made for capillary pressure. Note

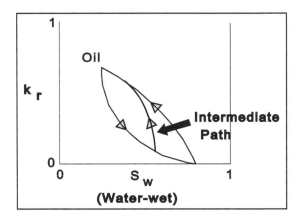

Relative permeability intermediate path

that hysteresis effects may occur due to changing injection patterns, infill drilling or simply shutting in wells. As would be expected, including these effects in a simulator may lead to the use of smaller timesteps or poor material balances.

Recent laboratory data indicates that many secondary drainage curves for water-wet systems actually lie below the imbibition curve for oil relative permeability.

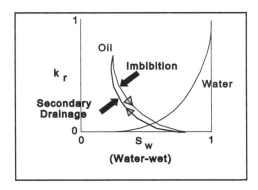

Recent secondary drainage data

To sum up hysteresis effects with respect to wettability: in water-wet systems, oil relative permeability may be affected (water relative permeability usually has a negligible effect); in oil-wet systems, water relative permeability may have a significant effect; finally, in intermediate or mixed wettability systems, all bets are off, meaning either one, both or none of the relative permeability curves may be affected.

Hysteresis effects may also occur in gas–liquid (either gas–oil or gas–water) relative permeabilities; in this case, the gas relative permeability can change from drainage to imbibition as it is the nonwetting phase. When undergoing imbibition, the increased critical gas saturation is usually referred to as the trapped gas saturation.

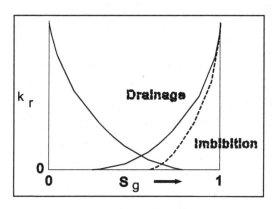

Gas–Oil hysteresis

When no hysteresis effects occur due to no flow reversals, select the appropriate relative permeability and capillary pressure data to simulate the flow process. See Secs. 3.5.2 (p. 26) and 4.5 (p. 38) of M-13.

Spacing and **smoothness** of saturation-dependent data will greatly affect the running time of a model. Relative permeability and capillary pressure data are usually not equally-spaced in simulators due primarily to the shape of the curves; also, since linear interpolation is used between data points, care must be exercised in selecting points that define the desired curve.

Quite frequently, extremely low values of relative permeability are added or adjusted to achieve proper phase productions;

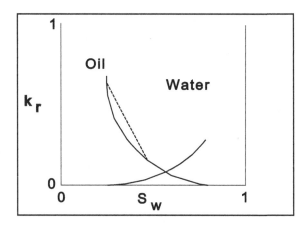

Interpolation effect

care must be taken to avoid creating a discontinuous curve and the best method available is to employ a semilog plot of $\log(k_r)$ as a function of saturation. As the complexity of a model increases (from IMPES to fully implicit), the importance of smoothed saturation dependent data increases.

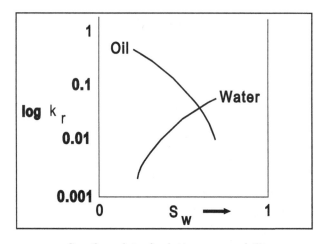

Semilog plot of relative permeability

When applying relative permeability data to zones with end-point saturation values which differ from the data available, adjusting the end-point values known as **end-point scaling**, is accomplished by using normalized saturations. In order to adjust the water–oil data, use a normalized water saturation

$$S_w^* = \frac{S_w - S_{wc}}{1 - S_{wc} - S_{orw}}$$

calculated using the original end-point $(S_{wc}$ and $S_{orw})$ values and then recalculate new values of S_w from S_w^* and the new end-points. For gas–oil systems, use a normalized gas saturation

$$S_g^* = \frac{S_g - S_{gc}}{1 - (S_{gc} + S_{lc})}$$

where $S_{lc} = S_{wc} + S_{org}$.

Similarly, when you have too many sets of relative permeability data (*e.g.* data from more zones than you can model), you must **average the relative permeability sets**. To do this, calculate normalized saturation values at the same value of relative permeabilty from each dataset. Next, at a fixed relative permeabilty value, average the normalized saturations. Finally, average the end-point values and use them to back-calculate saturations.

Weighting of relative permeability (often referred to as **mobility weighting**) data when used in a model affects the values employed to determine flow conditions; this is often referred to as transmissibility weighting. It is simply a technique of deciding which relative permeability values to use to determine flow into a block. Also, variations of standard upstream-weighting methods are merely another way of altering the relative permeability curves.

Consider two cells at a flood front as shown. The interface between the cells $(i + \frac{1}{2})$ will determine the flow capability from one cell to another. One idea would be to average the relative permeability values, which looks pretty reasonable until the

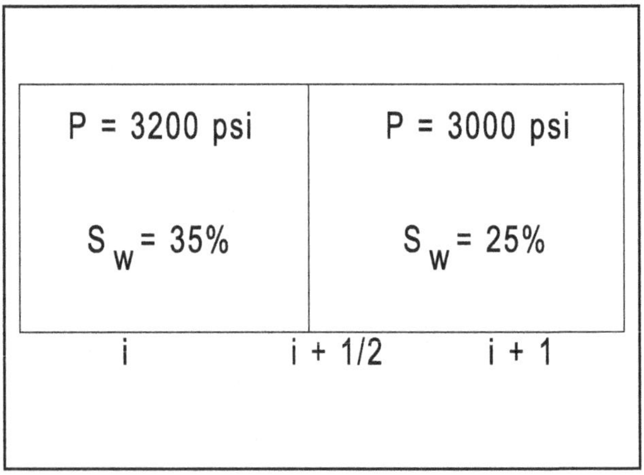

P = 3200 psi	P = 3000 psi
S_w = 35%	S_w = 25%
i i + 1/2	i + 1

situation shown below occurs in which case a non-zero water relative permeability value would remove critical water from the $i + 1$ cell. The most reasonable approximation is the use of **upstream-weighting** which uses the relative permeability value of the higher pressure (technically, higher potential) cell at the interface.

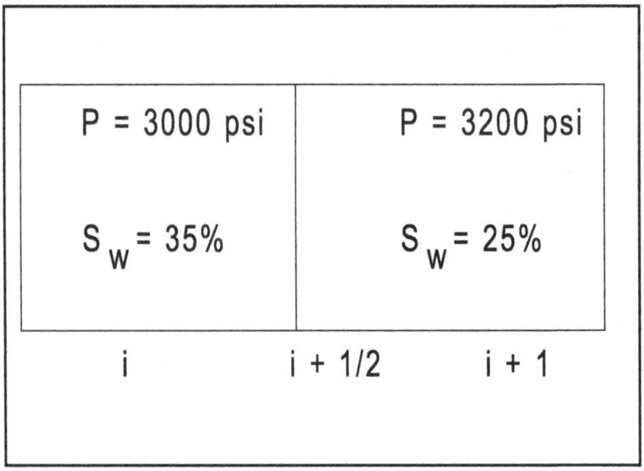

P = 3000 psi	P = 3200 psi
S_w = 35%	S_w = 25%
i i + 1/2	i + 1

An improvement on this technique which reduces gridding effects is two-point upstream-weighting although this variation requires additional computation time, particularly in fully

P = 3250 psi	P = 3200 psi	P = 3000 psi
S_w = 40%	S_w = 35%	S_w = 25%

i - 1   i - 1/2   i   i + 1/2   i + 1

implicit models. At the $i + \frac{1}{2}$ interface,

$$k_{rw_{i+\frac{1}{2}}} = \frac{3}{2}k_{rw_i} - \frac{1}{2}k_{rw_{i-1}}.$$

When a gradient is not consistent for two cells, regular (single-point) upstream-weighting is used. Additional discussions of mobility weighting may be found in Secs. 2.2.2 (p. 8), 5.5.1 (p. 52), 6.3.4 (p. 60) and C.1 (p. 141) of M-13.

Three-phase relative permeability values are sometimes required for oil since its flow may be restricted by both water and free gas. While this data may be measured in the laboratory, it is seldom available and a method of combining the two-phase data is usually employed. Most popular are the two methods developed by Stone; however, note that both of these correlations are based on data from water-wet systems. Other methods include various saturation-weighted interpolations of two-phase data.

The first method (Stone I, JPT February 1970) is based on a statistical analysis and is of the form

$$k_{ro} = S_{on}f_1f_2,$$

where

$$S_{on} = \frac{S_o - S_{or}}{1 - S_{or} - S_{wc}}$$

$$f_1 = \frac{k_{row}}{1 - S_{wn}}$$

$$S_{wn} = \frac{S_w - S_{wc}}{1 - S_{or} - S_{wc}}$$

$$f_2 = \frac{k_{rog}}{1 - S_{gn}}$$

$$S_{gn} = \frac{S_g}{1 - S_{or} - S_{wc}}.$$

Fayers and Matthews (SPEJ, April 1984) suggest that

$$S_{or} = W S_{orw} + (1 - W) S_{org},$$

where the weighting factor

$$W = 1 - \frac{S_g}{1 - S_{wc} - S_{org}}.$$

The second method of Stone (Stone II, JCPT 1973) is based on physical model results and is a combination of relative permeability values:

$$k_{ro} = (k_{row} + k_{rw})(k_{rog} + k_{rg}) - (k_{rw} + k_{rg}).$$

The main difference in these two models occurs at low k_{ro} values where the Stone II equation may yield negative values for k_{ro}. A preference has been shown for Stone II since it does not require a value for S_{or}; however, this author prefers Stone I.

Saturation-weighting to obtain three-phase oil relative permeability values has been advocated by various authors. One of the more popular methods is that of Baker (SPE 17369)

$$k_{ro} = \frac{(S_w - S_{wc})k_{row} + (S_g - S_{gc})k_{rog}}{(S_w - S_{wc}) + (S_g - S_{gc})}.$$

A great deal of concern has been expressed over the use of three-phase relative permeability equations and while use of various equations will yield different results, for reasonably gridded model studies, very few cells actually undergo three-phase flow. A survey of a number of model studies indicated that fewer than 5% of the cells actually required three-phase oil relative permeability unless exceptionally coarse gridding was employed (see Sec. 4.5.4, p. 40, M-13).

Pseudofunctions is a term that covers some reasonable variation from the norm for a specific purpose. Many types of pseudofunctions are available, but at present we will concern ourselves with variations in capillary pressure and relative permeability. Previously, we have already used a pseudofunction for capillary pressure when it was mentioned that for heterogeneous systems, the addition of several different sets of capillary pressure curves would approach a straight line. It was also pointed out that one of the primary uses of capillary pressure data was to determine the initial fluid (saturation) distributions. Knowing the saturations at a few locations, the capillary pressure curve may be back-calculated to obtain saturations throughout the reservoir. This is a simple type of a pseudofunction.

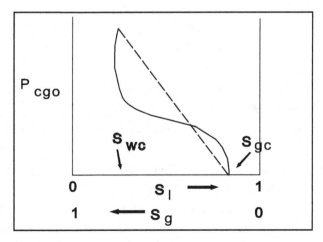

Capillary pressure pseudofunction

Pseudocapillary pressures can be generated when fluid distributions are known using

$$P_c = \frac{\Delta \rho H}{144}$$

as previously defined. For example, if the following water saturation data is known

$S_w(\%)$	Elevation (ft)	
30	-7100	(critical water)
50	-7150	
70	-7180	
100	-7200	

in a reservoir having a water density of $67\,\text{lb/ft}^3$ and an oil density of $50\,\text{lb/ft}^3$, the pseudocapillary pressures would be calculated from $P_c = (67 - 50)H/144$ where "H" is measured from the 100% water depth upward, resulting in

$S_w(\%)$	P_c (psi)
30	11.80
50	5.90
70	2.36
100	0.0.

Note that pseudocapillary pressures in this case would approach rock capillary pressures if very thin cells were used; the pseudo is required due to the use of an average saturation in each cell.

In a slightly different context, when an areal model is used, even if little or no capillarity exist, a pseudocapillary data set is needed to obtain the correct average saturations in each cell. Assuming an extremely sharp gas–oil contact (implying no capillarity) in the model shown, the average gas and oil saturations must be generated from pseudocapillary pressure data. For an oil having a density of $50\,\mathrm{lb/ft^3}$ and a gas density of $9\,\mathrm{lb/ft^3}$, the following capillary values are needed.

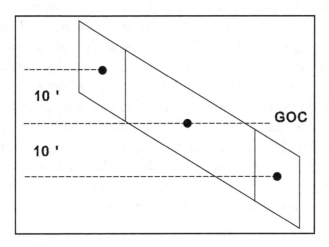

1. For the third cell, being all liquid (oil and critical water),

$$P_{cgo} = 0.0\,\mathrm{psi}.$$

2. For the middle cell, if $S_{wc} = 20\%$ and the GOC is in the center of the cell,

$$S_g = S_o = 0.5(1 - S_{wc}) = 40\%$$

and

$$P_{cgo} = \Delta\rho H/144$$
$$= (50 - 9)(10)/144$$
$$= 2.85\,\mathrm{psi}.$$

3. For the first cell, consisting of gas and critical water

$$S_g = 1 - S_{wc} = 80\%$$

and

$$P_{cgo} = (50 - 9)(20)/144$$
$$= 5.70\,\mathrm{psi}.$$

Obviously, more intermediate cells would define more capillary values.

Vertical equilibrium is another concept that uses pseudofunctions. Whereas the standard concept is one of diffused flow within any given cell, the VE concept attempts to account for some degree of segregation of fluids in a cell. Pseudocapillary pressures (although true capillarity does not exist) are calculated as shown previously, usually from known saturation profiles. The pseudorelative permeability data required for complete separation of fluids is a linear relationship between the end point relative permeability values. While the VE concept applies to gas–liquid systems where fine vertical gridding cannot be used, for water–oil systems, a very large vertical permeability would be required.

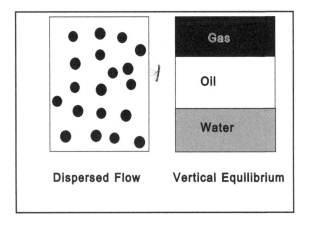

The concept of the VE shift in the relative permeability curves leads us to two important observations. First, if relative permeability curves are adjusted, the upper limit (for normal reservoir flow conditions) is a straight line. Second, as we relax the relative permeability curve, we imply a certain degree of fluid separation as shown in Fig. 3.22, p. 21, M-13. Many VE options allow mixing of the VE data with rock data or use of VE

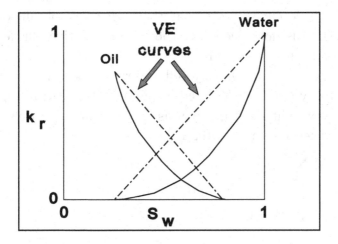

in certain directions (e.g. to retard vertical water movement). A discussion of vertical equilibrium is in Sec. 3.3.2, p. 21, M-13.

Many variations of pseudorelative permeability are for **simulating flow in stratified reservoirs** when using an areal model or reducing the number of layers to be modeled. A simple version of this application is to employ end-point values to determine maximum allowable relative permeability curves. This technique (attributed to Hearn) will usually cause very little change in the non-wetting phase relative permeability, but will yield a convex

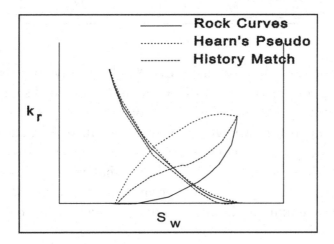

wetting phase curve. Note that the final curve obtained using this method is not necessarily the correct relative permeability data, but an upper bound.

For example, in the layered reservoir shown, the average permeability can be calculated using an arithmetic average permeability weighted by net pay as detailed in the absolute permeability section in Chapter 1.

$$k_{avg} = \sum (kh)/\sum (h)$$
$$= [(50)(10) + (10)(10) + (100)(10)]/(10 + 10 + 10)$$
$$= 53.33 \, md.$$

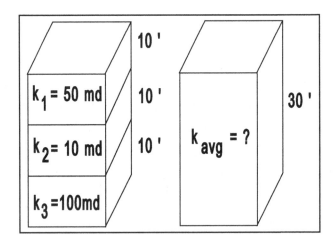

Use of normal laboratory measured "rock" relative permeability curves in conjunction with the arithmetic average absolute permeability requires that the saturation in each layer be approximately equal the average cell saturation. In reality, the above displacement must therefore occur with sufficient crossflow to
. equalize the frontal velocities (a somewhat restrictive assumption). Hearn's pseudorelative permeability concept allows us to develop maximum-bounding relative permeability curves for

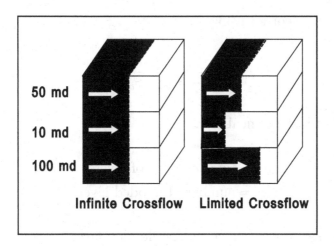

use in an areal model with an arithmetic-average absolute permeability to simulate stratified flow. It assumes unit mobility, non-communicating layers and piston-like displacement.

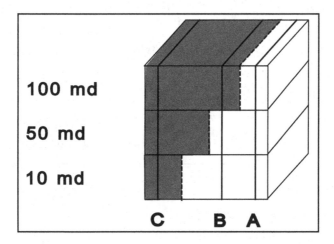

This concept requires that we reorder the layers in order of decreasing permeabilities (as in Stiles method) so that the relative permeability data may be calculated in order of increasing water saturation; for varying porosities and end-point saturations, reorder using

$$\frac{k}{\phi\left(1 - S_{wc} - S_{orw}\right)}$$

instead of just permeability.

$$nz = \text{total number of layers.}$$
$$n = \text{number of flooded layers.}$$

At cross-section A (which has no flooded layers):

$$S_w = \frac{\sum\limits_{i=1}^{nz} h_i \phi_i S_{wci}}{\sum\limits_{i=1}^{nz} h_i \phi_i}$$
$$k_{rw} = 0$$
$$k_{row} = k_{row}@S_{wc}.$$

At cross-section B:

$$S_w = \frac{\sum\limits_{i=1}^{n} h_i \phi_i (1 - S_{orwi}) + \sum\limits_{i=n+1}^{nz} h_i \phi_i S_{wci}}{\sum\limits_{i=1}^{nz} h_i \phi_i}$$

$$k_{rw} = \frac{k_{rw} \sum\limits_{i=1}^{n} k_i h_i}{\sum\limits_{i=1}^{nz} k_i h_i}$$

$$k_{row} = \frac{k_{row} \sum\limits_{i=n+1}^{nz} k_i h_i}{\sum\limits_{i=1}^{nz} k_i h_i}.$$

At cross-section C:

$$S_w = \frac{\sum\limits_{i=1}^{nz} h_i \phi_i (1 - S_{orwi})}{\sum\limits_{i=1}^{nz} h_i \phi_i}$$

$$k_{rw} = k_{rw} @ S_{orw}$$

$$k_{row} = 0.$$

For the example shown:

$$\phi \,(\text{all layers}) = 14\%$$
$$S_{wc} \,(\text{all layers}) = 0.168$$
$$S_{orw} = 0.375;\, 1 - S_{orw} = 0.625$$
$$\sum kh = 1600\,\text{md} - \text{ft}$$
$$\sum \phi h = 4.2\,\text{ft}$$
$$k_{rw} @ S_{orw} = 0.55$$
$$k_{row} @ S_{wc} = 0.92.$$

This data will generate $(nz + 1)$ points.

At cross-section A (no layers flooded):

$$S_w = 0.168$$
$$k_{rw} = 0$$
$$k_{row} = 0.92.$$

At cross-section B (layer 1 has been flooded):

$$
\begin{aligned}
S_w &= [(10)(0.14)(0.625) + (10)(0.14)(0.168) \\
&\quad + (10)(0.14)(0.168)]/4.2 \\
&= 0.32 \\
k_{rw} &= (0.55)(100)(10)/1600 \\
&= 0.344 \\
k_{row} &= (0.92)[(10)(10) + (50)(10)]/1600 \\
&= 0.345.
\end{aligned}
$$

At a point half-way between cross-sections B and C (layers 1 and 2 flooded):

$$S_w = [(10)(0.14)(0.625) + (10)(0.14)(0.625)$$
$$+ (10)(0.14)(0.168)]/4.2$$
$$= 0.473$$
$$k_{rw} = (0.55)[(100)(10) + (10)(50)]/1600$$
$$= 0.516$$
$$k_{row} = (0.92)(10)(10)/1600$$
$$= 0.0575.$$

At cross-section C (all layers flooded):

$$S_w = 0.625$$
$$k_{rw} = 0.55$$
$$k_{row} = 0.$$

If this technique is employed using saturations and relative permeability values from a cross-sectional model at various times (instead of end-point values), it yields dynamic pseudofunctions. This technique could also be used when different relative permeability tables existed (as for varying connate water saturations), but it becomes somewhat self-defeating by oversimplifying the problem (i.e., a 3-D model should be used).

Dynamic pseudofunctions is a term applied to relative permeabilities and capillary pressures determined from cross-sectional simulators for use in areal models. This method is used in lieu of a large 3-D model. These pseudofunctions will vary with both location and time, and as such require great care in their use. Also, data must be modified to account for grid variations between the cross-sectional and areal models. Essentially, one or more cross-sectional model studies must be made for each areal study and various sets of relative permeability data are input to the areal model depending on the fluid distributions at any point in time as shown in Figs. 3.24 and 3.25 (pp. 22–23, M-13). A detailed explanation of dynamic pseudofunctions may be found in Sec. 3.3.3, p. 22, M-13; note that there are no recent references to this method.

A slightly different twist is to use a radial model to study individual well traits (such as coning) and develop pseudorelative permeability relationships for well cells as illustrated in Figs. 7.8 and 7.10, p. 81, M-13 (see Sec. 7.4.1.8, p. 82, M-13). The coning productions from the radial model are used to back-calculate relative permeability values which are activated in the cartesian model well cells when the critical coning rate is exceeded.

Simulator requirements of relative permeability data vary; some simulators require slightly negative saturation values to prevent numerical overshoot. One important consideration is the expansion of critical water due to pressure drop. As the pressure drops, the critical water saturation expands slightly; however, in the relative permeability data, at any value above critical, a linear interpolation is used and as such a small value is calculated for the water relative permeability. Unless there is a definite reason to produce very small amounts of water, the simulator solutions will be less time-consuming for two-phase flow than for oil, gas and water.

The addition of an irreducible water saturation slightly $(0.5 \rightarrow 1\%)$ above critical having $k_{rw} = 0$ and k_{row} slightly less than the value at critical water will prevent water flow due to expansion.

Multiple sets of relative permeability tables are employed when end-point values (e.g. critical water saturation) vary within a field (stratification effects). Many models allow simply shifting an existing relative permeability curve to account for the varying critical (or connate) water saturations; this technique is known as **end-point scaling** and has been discussed earlier in this chapter. Multiple sets of capillary pressure tables may also be needed when porosity and permeability vary (as previously discussed in Chapter 1 under J-functions). Additionally, changing wettability or the use of pseudofunctions (e.g. coning) will require different sets of tables. When multiple tables are used, each table is assigned to a group (or groups) of cells and fluids moving from one table assignment to another are instantly governed by the new table. See Secs. 4.5 (p. 38) and 4.5.4 (p. 40) of M-13 for additional information.

Some models allow input of different relative permeability tables in the vertical and areal directions as an option.

Problems — Chapter 4

1. *End-point scaling of relative permeability.* Calculate the appropriate water saturation values to scale the given water–oil relative permeability data for the following end-point values

$$S_{wc} = 15\%$$
$$S_{orw} = 40\%$$

from the original data shown below

S_w	k_{row}	k_{rw}
0.104	1	0
0.212	0.64	0.019
0.283	0.38	0.036
0.427	0.058	0.058
0.48	0.01	0.067
0.57	0	0.09

2. *Averaging of relative permeability.* Calculate the average water–oil relative permeability data for the following three

datasets:

1.

S_w	k_{row}	k_{rw}
0.10	1	0
0.25	0.27	0.005
0.40	0.06	0.022
0.55	0.0043	0.080
0.60	0.001	0.125
0.72	0	0.35

2.

S_w	k_{row}	k_{rw}
0.20	1	0
0.30	0.26	0.007
0.50	0.06	0.035
0.60	0.013	0.092
0.70	0.01	0.23
0.74	0	0.35

3.

S_w	k_{row}	k_{rw}
0.30	1	0
0.40	0.45	0.0043
0.50	0.15	0.016
0.60	0.048	0.05
0.70	0.008	0.15
0.78	0	0.35

3. *3-phase relative permeability.* Calculate the oil relative permeability values using both techniques of Stone for the following 3-phase flow conditions:

$$S_{wc} = 25\%$$
$$S_{orw} = 20\%$$
$$S_{org} = 10\%$$

	Case 1	Case 2	Case 3
S_w (%)	30	60	30
S_o (%)	30	30	50
S_g (%)	40	10	20
S_{or} (%)	13.85	18.46	16.92
k_{row}	0.6	0.1	0.6
k_{rog}	0.2	0.7	0.6
k_{rw}	0.1	0.4	0.1
k_{rg}	0.25	0.05	0.1

S_{or} values were calculated using the correlations of Fayers & Matthews.

4. *Pseudocapillary pressures.* Calculate the gas–oil capillary pressures (as a function of either gas or liquid saturation) for the following reservoir:

API gravity $= 45.1°$ gas gravity $= 0.65$

$B_o = 1.18$ RVB/STB $B_g = 2.5$ RVB/MCF

$R_s = 480$ SCF/STB $S_{wc} = 25\%$

S_g (%)	Elevation (ft ss)
75	-5420
50	-5424
25	-5426
0	-5428

5. *Pseudorelative permeability — Stratified system.* Calculate the water and oil pseudorelative permeability values for a stratified reservoir having a critical water saturation of 20% and a residual oil (to water) saturation of 25% in all layers. The oil relative permeability at the critical water saturation is 0.85 and the water relative permeability at residual oil is 0.40.

Layer	Permeability (md)	Thickness (ft)	Porosity (%)
1	40	4	18
2	5	6	14
3	80	5	15
4	10	12	8
5	150	7	16

Chapter 5

TRANSMISSIBILITIES

The term transmissibility is actually used in two very similar contexts in models. As a stand-alone term, it becomes a "holding" variable for a number of variables which normally do not change with time. When preceded by a phase connotation (oil, gas or water), it refers to the "holding" variable multiplied by the appropriate reservoir mobility.

Transmissibility equations are not extremely complicated; however, locating the correct values in the grid for modifications can be confusing. We can modify transmissibilities to simulate faults, shales, large frac jobs, and irregular grids. For flow in the

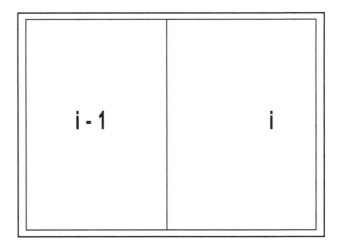

x-direction transmissibility

x-direction, with somewhat similar rectangularly shaped cells, the transmissibility is

$$AX_i = C \frac{k_{i-1}k_i \Delta y \Delta z}{(k_i \Delta x_{i-1}) + (k_{i-1} \Delta x_i)}$$

and for this example, the transmissibility exists at the interface between the "$i-1$" and the "i" cell. "C" is the appropriate conversion constant and includes the value of $1/2$ used in averaging in the denominator. For transmissibilities in the x-direction the numbering sequence begins at the top left and moves to the right; after completing one row, the sequence goes to the next row, etc. Note that there will be one more transmissibility per row than there are cells, and that the left and right end transmissibilities must be zero to avoid flow out of the grid (don't worry about this, the models handle it automatically). For simulators that show only NX transmissibilities, usually the right-end value is omitted although it does exist internally.

Since the x-direction transmissibility consists of two multiplicative permeabilities times the cross-sectional area to flow, all

Areal transmissibilities

divided by a length-weighted average permeability, the transmissibility in the y-direction is easily obtained using a similar logic.

$$AY_j = C\frac{k_{j-1}k_j\Delta x\Delta z}{(k_j\Delta y_{j-1}) + (k_{j-1}\Delta y_j)},$$

where the permeabilities now represent y-directional permeabilities. There will be $NY + 1$ transmissibilities with the front and back set to zero to seal flow from the grid.

y-direction transmissibilities are numbered from rear to front, so that using a simple 2×2 grid we have the following transmissibilities available for modification:

Interface	i	j	Trans.
A	2	1	AX
B	2	2	AX
C	1	2	AY
D	2	2	AY

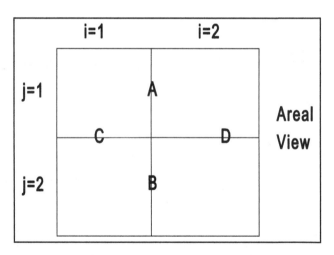

Areal transmissibility example

To thoroughly confuse the issue, we must also deal with vertical or z-direction transmissibilities

$$AZ_k = C\frac{k_{k-1}k_k\Delta x\Delta y}{(k_k\Delta z_{k-1}) + (k_{k-1}\Delta z_k)},$$

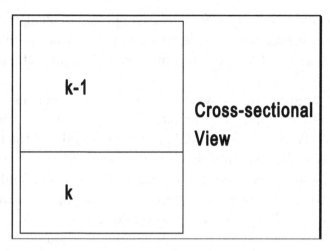

Vertical transmissibility

where the permeabilities are vertical permeabilities. Numbering occurs from top to bottom with the top and bottom layers set to zero. Since areal models have only a top and bottom (with regard to vertical transmissibility), no vertical flow exists and no vertical permeabilities are required. Obviously, a similar argument applies to AY values in cross-sections. Finally, although not shown, the appropriate interface may be found using averaging techniques when employing cell-variant thicknesses in an areal model.

Transmissibility values may be manually determined but it is much simpler to have the simulator internally calculate them. For unusual grid configurations such as a non-orthogonal or

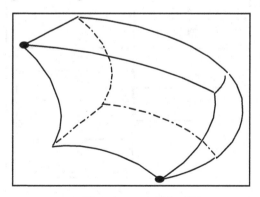

Curvilinear grid cell

curvilinear shaped grid, manual transmissibility calculations or a preprocessor program may be required (see Corner Point Gridding in Chapter 6); take my word for it, you want to use the preprocessor.

The majority of the work required using transmissibilities consists of modifying the simulator-calculated (or preprocessor-calculated) values. These modifications are usually in the form of a multiplier although additive or replacement values are allowed in many simulators. Once a transmissibility has been multiplied by zero (to create a sealing condition), multiplying it by any other values will not allow flow. When using additive or replacement values, be extremely careful to avoid changing the boundary values.

Fault simulation is one of the simplest uses of transmissibility modifications. If it is possible to align the grid with the fault, only one directional transmissibility may require modification. A multiplier of zero yields a sealing fault and values between 0 and 1 give varying degrees of leaky faults. Usually, a combination of x- and y-direction transmissibilities will require modification. Note that any multipliers less than one indicate a flow restriction (compared to the original conditions). Fig. 5.3 (p. 45, M-13) shows fault modeling.

Non-neighbor or **non-local connections** between cells may be employed for leaky faults or other types of formation

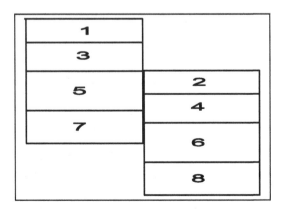

Non-neighbor cell connections

shifting. In the figure shown, cell 2 has a horizontal flow connection with cell 5; one-half of the horizontal flow from cell 4 would enter cell 5 and the other half would enter cell 7. Similarly, a reservoir may consist of formations with varying vertical flow connections and the use of non-neighbor connections would greatly reduce the number of zero pore volume cells required. Note that using non-neighbor connections will require substantial additional input data or, preferably, a preprocessing program.

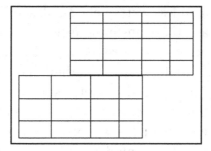

Non-neighbor vertical connections

Shales are modeled similar to faults with the exception that shales require a modification of vertical transmissibility.

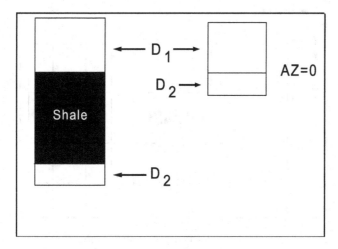

Shale modeling

By sealing vertical transmissibilities and proper input of cell elevations, thick shales may be simulated without requiring any actual cells for the shale. Similarly, multiple, non-communicating reservoirs with common completions may be modeled simultaneously by zeroing all vertical transmissibilities.

Large fracture treatments to a well may be simulated by increasing the transmissibilities (usually AX and AY) of the well cell. This will allow greater flow into the well cell; small frac jobs should normally be simulated by adjusting the skin factor of the well unless extremely small well cells are being used (in which case the transmissibilities should be modified). Fracture treatments may require modifying transmissibilities at a time other than initialization. Note that a doubling of reservoir permeability in a given cell does not result in a doubling of the transmissibility (unless adjoining cells are similarly affected).

Naturally Fractured Reservoirs are **not** correctly modeled using transmissibility modifiers. For example, a fracture lying in the x-direction as shown cannot be modeled by increasing the y-direction transmissibilities since they control flow in the y-direction and do not channel flow in the x-direction.

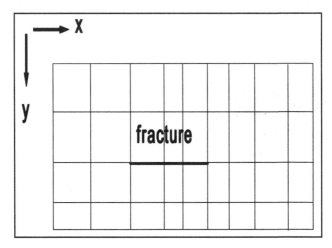

Incorrect fracture model using transmissibility modification

Specific fracture flow of this type should be studied using a **dual porosity** model (as explained in Chapter 9). Naturally fractured reservoir modeling is discussed in Sec. 3.3.5, p. 24, M-13.

Problems — Chapter 5

1. *Transmissibility modifications.* Indicate the locations and transmissibilities to be modified to model the sealing faults shown.

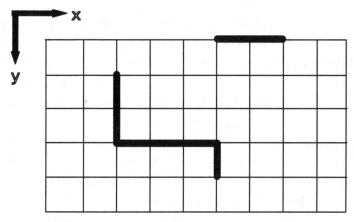

Location   Multipliers
I   J   AX   AY

Chapter 6

GRIDDING CONSIDERATIONS

The field to be studied must be overlaid with a grid to define each section of the reservoir(s). Prior to gridding the model, the correct type of simulator must be selected. The importance of poorly-defined parameters should be determined by using simple models. Also, the effect of the grid selected may be found by running the model using a finer grid to see if any significant changes occur; these techniques of varying model parameters to determine the effect on model results are known as **sensitivity testing**. They should be investigated prior to the actual study, but are usually carried out after a study is completed.

Fluid dispersion is assumed to exist in every cell as shown in Fig. 2.6, p. 10, M-13. If special concepts (such as vertical equilibrium) are employed, they adjust the data to yield the desired results based on a dispersed system. Shifting of relative permeability curves to prevent early fluid movement is required due to the dispersed fluid concept. The reservoir shown could be modeled with a coarse areal grid (all cells of equal elevation) but extreme data adjustment would be required to match actual field performance. A better representation would be to use a cross-section with layers defined by the fluid contacts. Depending on the amount of areal variation, a 3-D model may be required in lieu of a cross-section. Note that while an edge-water aquifer would simplify use of an areal model with varying elevations, most cells would still have two-phase dispersion.

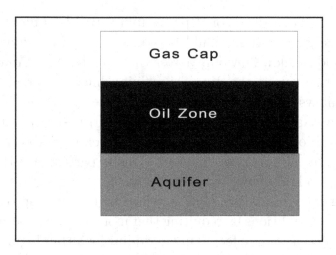

Fluid contact grid

With these concepts in mind, we may formulate some gridding considerations.

Heuristic gridding rules are a combination of experience and common sense. Some gridding selection concepts in order of consideration are shown below.

Areal Grids

- Anisotropy
- Flow via diagonal cells
- Rotate grid to minimize empty cells
- Locate axes parallel to faults
- Minimum of three cells between wells
- Center wells in cells
- Plan ahead for future wells
- Avoid large-to-small neighboring cells

Cross-sectional Grids

- Geologically varying strata
- Existing well completions
- Future well completions
- GOC & WOC

When strong directional permeability trends (**anisotropy**) exist, the grid system should be parallel to these trends since most models allow flow only across the cell interfaces. An exception to this rule is the nine-point finite difference areal model which allows flow from octagonal-shaped cells.

Flow via diagonals (from well to well) best approximates field conditions (requiring the least amount of relative permeability adjustment to match history). This is particularly true for injectors and producers; whenever possible, select a grid to avoid aligning injection and production wells. This tends to approximate radial flow conditions by requiring that more interfaces be crossed (essentially the same effect as finer gridding for all but highly unfavorable mobilities). See Figs. 5.17, 5.18, and 5.19, pp. 53–54, M-13. Grid effects are minimized by using two-point upstream-weighting or a nine-point flow system as shown in Fig. 5.20, p. 55, M-13 (which has not become popular). The nine-point non-orthogonal system gives increased accuracy for distorted grids (note that it is nine-point in an areal mode, 27-point in 3-D).

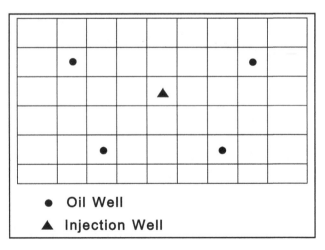

Areal gridding example

Next, rotate your grid (either areal or cross-sectional) to minimize the number of cells not in the simulated region; these cells have no porosity (from a modeling viewpoint) and are referred to as "zero pore volume cells", "void cells" or "null cells". While they require no mathematical calculations, they may require computer storage and must be checked at several points during each timestep for elimination from the calculations. Note that from a modeling concern, it does not matter whether we run out of porosity, pay or cells (unless we plan to drill or fracture through a barrier of some type into additional reservoir). Areal grids need not run N-S and E-W! So, **rotate your grid to minimize the number of zero pore volume cells** whenever possible.

In orienting the grid system, if **faults** are **parallel to the axes**, it is much simpler to define them. Do not forget that the edges of the model are no-flow boundaries. If possible, use this fact to position one model edge along a sealing fault.

If individual well data is important (particularly in certain portions of the reservoir), **a minimum of three cells should exist between wells**. This rule holds true for both areal and cross-sectional studies (as well as 3-D). When studying a certain portion of a large field, honor this rule in the areas of interest, but do not finely grid areas of little concern or poor data as shown in Fig. 5.4, p. 46, M-13. Multiple wells may be located in the same cell, but unless they have similar completions and flow conditions, it is fruitless to attempt even an overall match with actual performance.

Wells should be located as close as possible to the **center of a cell**; results will not vary significantly as long as the well is in the central one-half of the pore volume of the cell.

A common sense rule, but one often overlooked, is to **plan ahead for future wells** when laying out the original grid.

In general, **surrounding cells** should not vary in pore volume by more than a factor of two (halving or doubling) for ease of timestep solution. This rule may be loosened to a factor of five in some cases. It is primarily designed for IMPES simulators although failure to follow it will lead to saturation smearing in fully implicit models. An important concept related to cell sizing is that of pore volume throughput (i.e., how much of a cell's pore volume can be changed in a given timestep). This is usually expressed as a maximum size of timestep where

$$\Delta t_{PV} = \frac{PV}{Q_o B_o + Q_w B_w + Q_g B_g}$$

and applies primarily to well cells. For studies undergoing rapid saturation changes, a maximum throughput of 15% appears reasonable; as saturation fronts stabilize (e.g., late in a waterflood), the throughput may be increased in the range of 60–70%. For IMPES simulators, this implies a danger of placing wells in extremely small cells.

Gridding **cross-sectional** models is usually easier to define than areal considerations. Of immediate concern in defining model layers are geologic stratification effects (permeability, porosity, connate water saturations, shales, etc.). As previously explained, as long as the correct elevations are maintained, non-productive intervals need not be included. Well completions will obviously also determine grid layers, and again, plan for future completions. Next in consideration are the gas–oil and water–oil contacts. Coarse vertical gridding will result in saturation smearing (see Fig. 5.10, p. 48, M-13) or lack of proper fluid movement (see Fig. 5.11, p. 48, M-13). Thick formations should avoid the use of thick cells to minimize fluid dispersion effects.

A **3-D model** employs both the areal and cross-sectional selection criteria presented, usually in that order.

Modeling an aquifer presents a special problem in that the size of the aquifer is usually huge when compared to the reservoir. A good technique for gridding the aquifer is to use normal gridding considerations to reach the aquifer, then use at least two areal cells doubling in size, and finally a third doubling cell. If the size of the aquifer must be altered, modify only the pore volume of the large (last) cell. Using an exceptionally

long length will invalidate the transmissibility equation which
we used in Chapter 5 (for cartesian systems):

$$AX = C \frac{k_{i-1} k_i \Delta y \Delta z}{k_{i-1} \Delta x_i + k_i \Delta x_{i-1}}$$

and if $\Delta x_i \gg \Delta x_{i-1}$, very little flow would occur due to the
large value in the denominator. It is interesting to note that for
a strong water drive, it is practically impossible to make the
aquifer too large. For strong water drives a water injection well
may be used in place of a large pore volume cell. Aquifers may
also be modeled with boundary modifications for special aquifer
functions such as constant flux or the Carter–Tracy (a modification of the van Everdingen–Hurst constant terminal rate case) or
Fetkovich (a pseudosteady state productivity index calculation
suitable for smaller aquifers) methods.

No discussion of gridding would be complete without some
comments concerning **numerical dispersion** (incorrect fluid
movement due to the gridblock effect). Figure 2.3, p. 8, M-13
shows a simple four cell model with a water saturation profile. In the next timestep, water will flow into Block 3 and then
after another timestep, some water will flow into Block 4 and
be produced. Use of more cells would require more timesteps
until water production begins, but requires exceptionally fine
grids. Judicious application of the guidelines given in this chapter coupled with previously discussed topics such as two-point
upstream weighting and pseudofunctions will yield adequate
results. Numerical dispersion is corrected when matching historical reservoir performance by adjusting the relative permeability data; when no historical data is available, use a fine grid
for a portion (or all) of the reservoir and match the fine grid
performance in the coarser grid (again by adjusting relative permeability data). Use of large timesteps (implying large pressure
and saturation changes) would result in another type of numerical dispersion. Additional information concerning numerical

dispersion may be found in Secs. 2.2.3 (p. 8), 5.5 (p. 51) and 6.4 (p. 60) of M-13.

Finally, we must be concerned with **partial cells** due to the fact that unless a very fine grid is employed, several cells will contain only part of the reservoir as shown in Fig. 5.2, p. 45, M-13. For example, if we are going to study the field shown using an areal model, we could adequately grid the field as illustrated; however, a number of our cells have only a partial pore volume

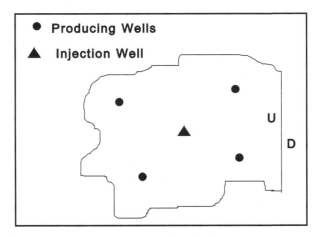

Field to be gridded

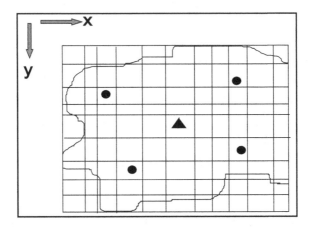

Field with areal grid

(i.e., the reservoir does not fill the cell). In this case, we have two possibilities: (1) either add pieces from cells with very little reservoir to other cells, or (2) use a reduced pore volume in the cells.

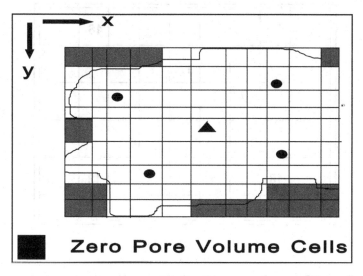

Final grid

A satisfactory approach is to **add on cells with less than one-third pore volume and adjust pore volumes for the remaining cells**. It is strongly recommended that pore volume multipliers be used for this purpose and that the actual porosity values remain unchanged. Note that simply reducing the pore volume in every cell could result in extremely small pore volumes next to large pore volumes, a situation that results in saturation smearing and severe timestep limitations as previously discussed in this chapter.

Cell property determination must be calculated to fit the grid. Once the grid has been established, porosities, elevations, directional permeabilities, and thicknesses must be assigned to each cell; saturations and pressures may also be required. Assuming we had selected the grid system with porosity values shown in the well cells (these values would be available from logs, core analysis or correlations), the contour lines for an iso-vol map could be determined as shown, and by overlaying the grid pattern, a porosity value for each cell is obtained.

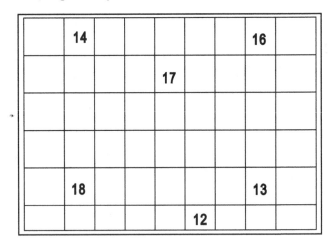

Areal grid with known values

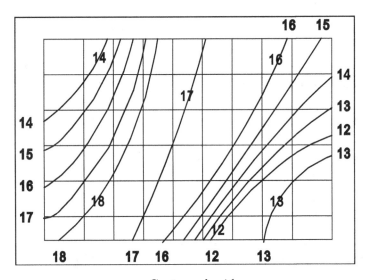

Contoured grid

Computer contouring programs are available for this purpose; however, these programs are not one-step solutions. They require frequent manual intervention (usually the addition of

14	14	16	17.8	17.6	16.9	16.4	16	15.1
13.7	15.1	16.8	17.7	17	16.7	16.2	15.6	14.2
14.8	16.2	17.4	17.5	16.8	16.5	15.7	14.3	12.9
15.7	17	17.8	17.3	16.6	16.2	14.2	12.4	12.8
17.1	18	17.5	16.8	16.2	14.5	12.4	13	13
18.2	17.8	17.4	16.6	15.5	12	12.5	13	13

Final porosity cell values

more control points) until a reasonable result is achieved. The variable most likely to require repeated adjustment is elevation; next is net pay.

Partial field modeling is accomplished by modifying transmissibilities, pore volumes and rates. In the example

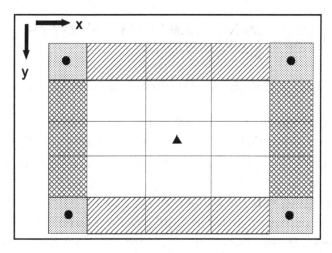

Partial field modeling

shown, no flow enters or leaves the pattern. Using the following definitions,

$$APV = \text{actual pore volume;}$$
$$AX, AY, AZ = \text{actual transmissibilities;}$$
$$AQ = \text{actual rate;}$$
$$MAX, MAY, MAZ = \text{modified transmissibilities}$$
$$\text{(to be used in the model);}$$
$$MPV = \text{modified pore volume;}$$
$$MQ = \text{modified rate,}$$

the necessary modifications are as shown in the diagram below.

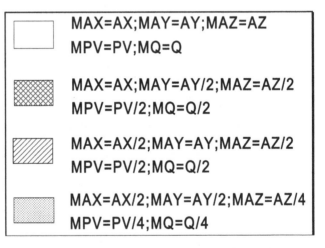

Modifications for partial field modeling

When no-flow conditions cannot be established, fine gridding should be used in the area of interest and coarse gridding for the remainder of the field. An example of this technique is to place thick slices on each side of a cross-section of interest.

Local grid refinement (LGR) allows the use of a fine grid within a coarse grid, as for example, defining well cells in a field-wide model. Its use increases computer time, but results in more

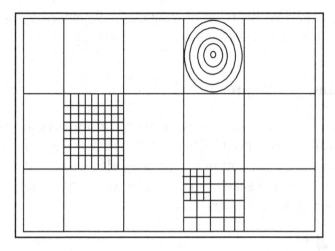

Local grid refinement

accurate simulations. For reasonable timesteps, a fully implicit formulation is preferred for the finer grid systems. The grid refinement may be either cartesian or radial within a cartesian system. Various other types of grids are currently under investigation, but the cartesian and radial grids remain the standards.

Corner point gridding allows the use of non-orthogonal shapes in describing reservoirs. It requires 8 locations for each

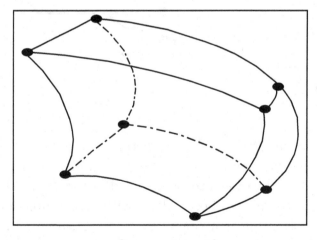

Corner point grid

cell and is not feasible to employ by manually entering grid data; as a result, a preprocessor program is needed.

Problems — Chapter 6

1. *Areal grid.* You are going to make an areal study of how water will influx into the field shown below. Set up a grid. How many cells did you use? You may also wish to study various water injection schemes to increase recovery. What changes would you make in the grid?

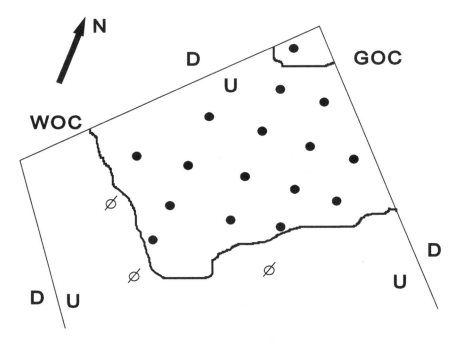

2. *Cross-sectional grid.* A representative cross-section of a field whose primary performance is to be studied is shown below. The SP logs for the five wells to be included in the model study are shown next to the wellbores along with the completion intervals. The formation is a coarse grained sandstone with higher permeabilities generally in the top 20′. A study of the shale

breaks (2′ to 20′ thick) indicates that they are not continuous over inter-well distances.

How many layers would you employ in your cross-sectional model? Draw the grid you would use. How many total cells comprise your model? How would you represent the shales in the model?

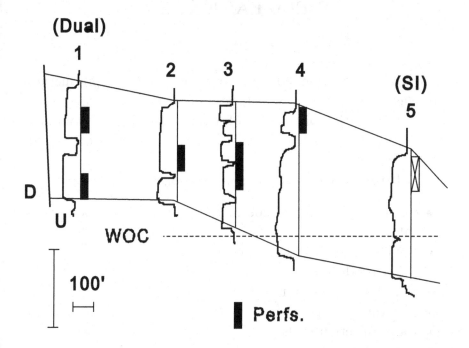

Chapter 7

WELL PACKAGES

Many different options are available for specifying production and injection in simulators. All these variations ultimately result in the specification of the rate of a single phase; the remaining two phases (if present) are then calculated and may be compared to observed data.

Well types based on rate specification include

- oil producers (oil rate specified, gas and water productions calculated),
- gas producers (gas rate specified, etc.),
- water producers,
- total liquid producers,
- reservoir voidage wells,
- water injection wells,
- gas injection wells.

For rate-specified wells, the given rate will remain constant (barring other controls) until manually changed. In matching past performance, we must use average rates over time periods when little fluctuation occurs. Future rates may be estimated using a decline curve analysis or productivity index calculation.

The remaining phase productions may be calculated using a mobility allocation scheme:

$$q_w = \frac{q_o B_o k_{rw} \mu_o}{B_w k_{ro} \mu_w}.$$

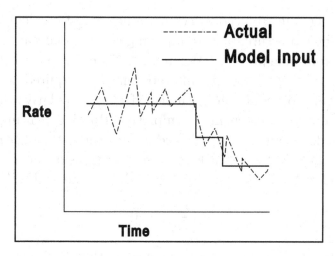

Rate averaging

When several layers are included in the well, layer productions must first be determined for the specified phase in each layer using a layer kh-mobility product ratioed to total kh-mobility product. The non-specified phases should be adjusted semi-implicitly after the new saturations are known in an IMPES model; in a fully implicit model, they are simultaneously calculated. The bottom-hole pressure for a well centered in a cell may be calculated (for isotropic conditions) using

$$P_{wf} = P_{\text{cell}} - \left(\ln \frac{r_{e\,\text{cell}}}{r_w} - 0.75 + s \right) \left(\frac{q_o B_o \mu_o}{k_{ro}} + \frac{q_w B_w \mu_w}{k_{rw}} + \frac{q_g B_g \mu_g}{k_{rg}} \right),$$

where, for reasonably sized cells with no directional permeabilities,

$$r_{e\,\text{cell}} = 0.14\sqrt{\Delta x^2 + \Delta y^2}.$$

When anisotropy exists,

$$r_{e\,\text{cell}} = 0.28 \, \frac{\sqrt{\Delta x^2 \sqrt{\frac{k_y}{k_x}} + \Delta y^2 \sqrt{\frac{k_x}{k_y}}}}{\sqrt[4]{\frac{k_y}{k_x}} + \sqrt[4]{\frac{k_x}{k_y}}}.$$

We have now introduced a technique not only for calculating a bottom-hole pressure, but for adjusting rates to maintain a specified bottom-hole pressure. When using the equation in the form of pressure control, small timesteps may be required to avoid an on-again, off-again jackhammer effect, particularly when an explicit production scheme is employed. Routine well improvements and/or damage are reflected in the skin factor or near-well permeability. Note that a lower limit for the skin factor in an improved wellbore (a negative value) due to the well cell size is

$$s = \ln \frac{r_{e\,\text{cell}}}{r_w} - 0.75$$

which makes $P_{wf} = P_{\text{cell}}$ and points out that the bottom-hole pressure for producers cannot be greater than cell pressure. When extremely small well cells are employed, a weight-averaged pressure, using the well cell and those surrounding it, should be used for P_{cell}.

A similar well specification used primarily for predictions is that of a productivity index (J or PI) which has units of bbl/day/psi. As the pressure drops, the rate drops correspondingly. The productivity index can be adjusted for the fact that the cell size is not representative of the actual drainage area by using a multiplying ratioing factor:

$$rf = \frac{\ln \frac{r_e}{r_w}}{\ln \frac{r_{e\,\text{cell}}}{r_w}}.$$

Efficiency factors can be employed to account for well downtime.

In a layered model, certain layers may be shown as closed to production (not perforated). Also, controls such as abandonment rate, gas–oil and water–oil ratio limits, and total field limits may be imposed. These all involve a simple check and adjust procedure. A desirable feature for many engineers is an automatic workover routine where layers are opened and closed to production based on GOR, WOR, backflow, etc. Care must

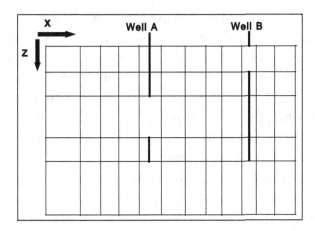

Closed layer model

be exercised when using these routines to avoid an unrealistic number of workovers. Some typical well controls are shown in Table 7.3, p. 76, M-13.

Pseudofunctions for wells may include any rate alterations based on conditions not normally associated with the type of model being used. For example, phase productions may be adjusted for perforated intervals in an areal model, or coning correlations may be included in cartesian coordinate models as shown in Figs. 7.10 and 7.11 on p. 81, M-13. Usually these types of correlations are developed using a radial simulator. When correlations such as these are employed, greater material balance discrepancies should be expected; also, timestep limitations may occur. Hopefully, a pseudofunction would not have to be derived for each well in the cartesian model, but wells may be grouped based on similar characteristics.

Model stability and timestep size depend to a large extent on the production-injection conditions employed. One example previously explained is the concept of pore volume throughput. Any time well rates are changed to any appreciable extent (including initial production), some consideration should be given to reducing the timestep, particularly if the number of solution iterations changes drastically. In a similar vein, large injectors may require a staged injection schedule, beginning at a lower rate and doubling the rate until total rate is obtained. Quite frequently, model difficulties due to well data are due to poor gridding considerations, too large of a timestep or data that would be physically impossible to implement in the field. Remember, the more well controls employed, the closer the model results should be screened.

Timestep size will be governed by saturation and pressure changes, desired output, required input (usually well changes), and numerical dispersion. Timesteps may be specified manually or automatically calculated by the simulator. When starting the model or after large well rate changes, use of a small timestep (1 day) is desirable to allow the model to "sort itself out." Use of extremely large timesteps will result in saturation (and pressure) smearing; however, the only way to determine that timesteps are too large is by rerunning the model using smaller timesteps to see if the answers change appreciably (sensitivity testing).

Automatic timesteps allow the user to vary the timestep size by employing an algorithm which uses saturation and pressure changes in conjunction with a minimum and/or maximum timestep size. When a model is running smoothly, this feature works well, but during rapidly changing portions of the study, manual intervention may be required, depending on the sophistication of the algorithm.

Rerunning a timestep using a smaller value when the solution does not meet the tolerances will reduce the amount of manual intervention required. A limited number of timestep reductions should be allowed so that a model does not continue to run when an obvious data bust is encountered (as a blown rate). Sometimes, reducing a timestep will not result in a good (or valid) solution; this is usually due to an accumulation of poor results which occurred previously in time; in other words, by the time you see it, it is too late.

Chapter 8

FIELD STUDIES

A **field study** consists of three primary phases: initialization, history match, and predictions. During **initialization**, be sure that the oil-in-place (and gas-in-place) is in the proper range. Also, at this point, check your saturation and pressure maps; of course, you have already checked your input data for errors (thickness, permeability, porosity, etc.). Scan the pore volume map to be certain that no small volume cells adjoin to large cells (many models now do this automatically). Transmissibility maps should be reviewed to assure correct placement of faults, etc. You are usually so eager to see the model run that initialization data errors are often overlooked. When both oil- and gas-in-place are in error, the first item to check is the gas–oil contact. Next, concentrate on oil-in-place.

All in-place values are simply volumetric calculations of the form (for oil)

$$\text{OIP} = \frac{S_o \, PV_o}{B_o}$$

and if a great discrepancy exists between the simulator and your estimate (or the accepted value), it must be in oil saturation, pore volume or oil formation volume factor. The latter is easily checked in the PVT table. Errors in oil saturation usually arise from incorrect water saturation values (check relative permeability/capillary pressure tables) or errors in elevation resulting in incorrect oil saturations. Once the oil-in-place is reasonable, it may be necessary to modify the entire reservoir using a global pore volume multiplier if a "prearranged" volume of oil is required. If OIP is correct, but the gas-in-place is in error, remember that it consists of both dissolved gas and free gas. First check the R_s values in the PVT table (a common error here is confusion between units: SCF/STB or MCF/STB), then check the terms involved in the free gas-in-place

$$\text{FGIP} = \frac{S_g \, PV_g}{B_g}$$

in a similar fashion to that just explained when checking oil-in-place. At this point, if you are satisfied with the data, the initialization is complete; this is the easiest part of the study. Remember, once the initialization is considered satisfactory, you will seldom return to it (basically, you have bought it at this point).

Once the initial description of the field is correct, you are ready to **match history**, or adjust your model description until it performs like the field did. Normally, you will specify the oil rate, so that the cumulative oil produced should always be correct. History matching parameters are

- **field pressure**,
- **water production** (usually on an individual well basis),
- **gas production** and
- **bottom hole flowing pressures** (if warranted by the data).

An exact match of past performance for all wells is seldom achieved. Usually, we hope to follow a pattern or trend of field performance. Remember that we normally consider relative permeability as our poorest data; however, if you have to adjust your relative permeability curve to a strange sinusoidal shape, there is something wrong with your description. Also, in a reservoir above the bubble point, gas production (or GOR) is not a valid matching parameter since all it indicates is that the solution gas entries in the PVT table are correct. The main purpose of history matching is to tune the model to perform like the field has in the past so that our future predictions will employ the best available data. A variety of reservoir descriptions may yield the same history match; however, of these descriptions, several will exceed acceptable limits of known parameters. A statement that this author often made is that given a set of reservoir data, this author can match history in up to two dozen model descriptions, but less than half a dozen will maintain reasonable reservoir modifications.

Production of "critical water" indicates that irreducible water saturation is set equal to critical in the relative permeability table; if this is not desired (or necessary), one can use a slightly higher irreducible value to increase the speed of the model.

A **systematic approach to history matching** is to

1. approximate non-specified phase productions,
2. verify field pressures,
3. fine-tune production and
4. fine tune bottom-hole pressures.

A simple method to check on the proper pressure level is to use reservoir voidage wells (although the phase productions will not be correct, the overall pressure level should be reasonable).

Adjustments required to approximate water and gas productions (for oil producers) are usually relative permeability modifications. Obviously, if the water production is too high, a reduction in the water relative permeability is needed at or before the water saturation in the well cell at that point in time.

History Match Example

A simple reservoir has been waterflooded using the relative permeability data as shown below. The actual WOR for the "B" well after 4 years was 1.0 while the model results (shown below) indicate a WOR of 2.93.

History Match Example

RELATIVE PERMEABILITY AND CAPILLARY
PRESSURE TABLE 1

S_w	$P_{c_{wo}}$	K_{rw}	K_{row}
0.3000	10.000	0.00000	0.43000
0.3100	9.857	0.00000	0.41800
0.4000	8.571	0.08500	0.31000
0.5050	7.071	0.19500	0.19500
0.6000	5.714	0.31000	0.12000
0.7000	4.286	0.44000	0.05500
0.8000	2.857	0.60000	0.00000
1.0000	0.000	1.00000	0.00000

CRITICAL WATER   30.00%
RESIDUAL OIL TO WATER   20.00%
IRREDUCIBLE WATER   31.00%

WELL ANALYSIS AT 4.00 (ELAPSED DAYS FROM START = 1460.00)

NO.		LOC. I	J	LYR TOP	BTM	CALC. BHP PSI	WATER RATE STB/ DAY	OIL RATE STB/ DAY	GAS RATE MCF/ DAY	GOR	WOR
OP	1 A	3	3	1	1	2154.0	5.9	96.4	614.4	6.371	0.06
OP	2 B	6	5	1	1	2353.0	224.9	76.7	344.7	4.495	2.93
WI	3 C	7	2	1	1	2621.0	300.0	0.0	0.0	0.000	0.00
TOTAL	(PRODUCTION)						230.8	173.1	959.1	5.540	1.33
	(INJECTION)						300.0		0.0		

OIL RECOVERY 20.16%

WATER SATURATION (%) AT 4.000 YEARS
(1460.00 DAYS)

K = 1

J	I = 1	2	3	4	5	6	7	8	9
1	30.4	30.4	30.4	32.9	41.6	48.8	53.1	33.9	30.9
2	30.4	30.4	31.7	36.7	52.5	63.3	71.1	46.5	32.2
3	30.4	30.4	31.6	35.1	50.8	61.8	64.3	43.3	31.8
4	30.4	30.4	30.6	31.8	46.8	57.9	56.7	39.3	30.6
5	30.4	30.4	30.4	30.4	38.8	46.5	46.7	33.4	30.4
6	30.4	30.4	30.4	30.4	30.4	31.6	32.8	30.4	30.4

Since the actual WOR was approximately 1/3 of the model WOR, the k_{rw} entries are reduced by a factor of 3 as shown below resulting in a reduction of the WOR for the "B" well from 2.93 to 1.50.

History Match Example — krel Adj.

RELATIVE PERMEABILITY AND CAPILLARY
PRESSURE TABLE 1

S_w	$P_{c_{wo}}$	K_{rw}	K_{row}
0.3000	10.000	0.00000	0.43000
0.3100	9.857	0.00000	0.41800
0.4000	8.571	0.02800	0.31000
0.5050	7.071	0.06500	0.19500
0.6000	5.714	0.10300	0.12000
0.7000	4.286	0.14600	0.05500
0.8000	2.857	0.20000	0.00000
1.0000	0.000	1.00000	0.00000

CRITICAL WATER	30.00%
RESIDUAL OIL TO WATER	20.00%
IRREDUCIBLE WATER	31.00%

WELL ANALYSIS AT 4.00 (ELAPSED DAYS FROM START = 1460.00)

NO.		LOC. I	J	LYR TOP	BTM	CALC. BHP PSI	WATER RATE STB/ DAY	OIL RATE STB/ DAY	GAS RATE MCF/ DAY	GOR	WOR
OP	1 A	3	3	1	1	2589.0	0.0	113.7	358.8	3.157	0.00
OP	2 B	6	5	1	1	2768.0	136.7	90.9	159.3	1.751	1.50
WI	3 C	7	2	1	1	3211.0	300.0	0.0	0.0	0.000	0.00
TOTAL	(PRODUCTION)						136.7	204.6	518.1	2.532	0.67
	(INJECTION)						300.0		0.0		

OIL RECOVERY 20.79%

WATER SATURATION (%) AT 4.000 YEARS (1460.00 DAYS)

K = 1

J	I = 1	2	3	4	5	6	7	8	9
1	30.3	30.3	30.3	31.2	46.0	59.8	67.7	36.7	30.7
2	30.3	30.3	30.3	37.2	62.4	69.9	74.5	56.7	34.1
3	30.3	30.3	30.6	34.0	59.5	69.5	71.6	50.3	31.4
4	30.3	30.3	30.3	31.1	51.0	66.3	64.5	39.7	30.3
5	30.3	30.3	30.3	30.3	34.5	52.1	54.7	30.4	30.3
6	30.3	30.3	30.3	30.3	30.3	30.9	33.9	30.3	30.3

Usually, a technique of bracketing the correct production is preferred because it indicates that the correct productions can be obtained by reasonable relative permeability adjustments (maximum upper limits to be considered are V. E. [straight lines], stratified systems or dynamic pseudofunctions). If water production cannot be bracketed by adjusting only the water relative permeability curve, the oil relative permeability curve must be reasonably adjusted in the opposite direction of k_{rw}. Failure to bracket water production at this point indicates the need for a review of the total reservoir description (both actual and grid data). Having approximated water production, gas production should be similarly bracketed. A semilog plot is useful for selecting low relative permeability values as shown

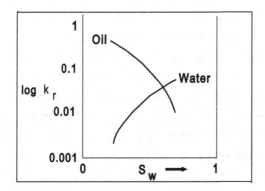

Semilog relative permeability plot

in Fig. 9.3, p. 104, M-13. The adjusted values must maintain a smooth curve. If these productions are reasonable, but over-all field pressure is not close to average pressure in the model, the pore volumes must be adjusted (this includes considering the presence and flow potential of an aquifer). If the average pressure is correct, but wellbore pressures or pressure trends are grossly in error, the absolute permeabilities should be adjusted (see Figs. 8.2 and 8.3, p. 91, M-13). Once again, failure to achieve reasonable results at this point warrants a review of the total reservoir description.

Average pressures calculated by the simulator may be either weighted pore volume (usually excluding high water saturation cells) or weighted hydrocarbon pore volume (HPV). HPV weighted cells may not reflect average pressures during water-flooding due to the reduction of oil and gas saturations. Average pressures may be calculated for the entire grid or for selected regions.

The remainder of the history match is now fine-tuning (first) the relative permeability data, and (second) the well data to match phase productions and bottom-hole pressures, respectively. In history matching, it is important to change only one variable at a time when performing bracketing; failure to follow this rule usually results in several changes which will not lead to a systematic approach, but rather a "hit-or-miss" with the usual result being a "miss." Note once again that there are several viable combinations that will yield a history match, so such a match is not unique but represents the most reasonable adjustments (usually to the poorest data) in the modeler's opinion.

Automatic history matching routines using regression analysis have not become extremely popular due to the complexity of reservoir descriptions and the computer time required. To use an automatic matching routine, you must first have a "reasonable" manual match (which is where most of your time is spent anyway). Chap. 8 of M-13 (p. 87) discusses various history matching items and shows some examples with varying history match parameters (e.g., Figs. 8.6 and 8.7, p. 93).

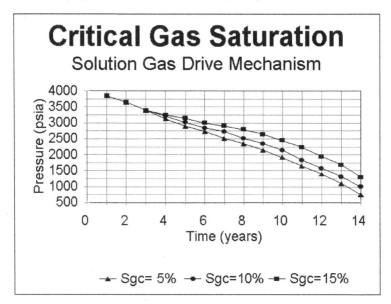

Critical gas sensitivity test

At this point (if not previously investigated), some **sensitivity testing** is in order. You will already have discovered in history matching that your results are more sensitive to certain variables than others. Parameters not greatly affecting performance need not be known to any extreme accuracy. Is the model

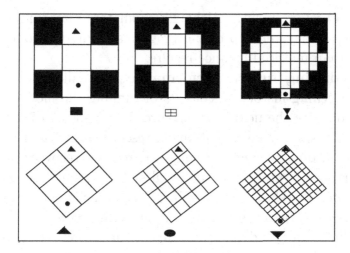

Grid sensitivity legend

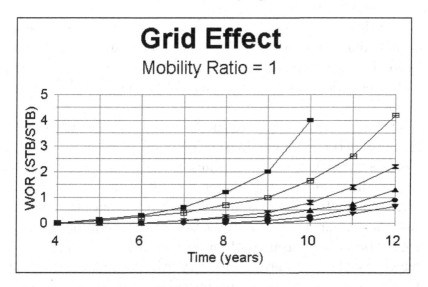

Grid sensitivity results

sensitive to gridding changes? This is easily tested although in actuality it usually involves considerable engineering time. Is the model sensitive to timestep size? All models are to some degree; however, if we have maintained reasonable saturation and pressure changes, our timestep selection should be satisfactory. Usually, a safe maximum timestep (when good history is available) is one month. Although simulators may run at large timesteps, they tend to over-average results and become little better than material balance results.

Predicting future performance is usually much easier than matching history; it consists of running the model as you would operate the field in the future. Your first predictive case should always be a continuation of present operations (i.e., what if we don't do anything?). If you initiate a new recovery process, as starting a waterflood, the results will be only as good as your water–oil relative permeability data (which possibly has not been used). You can get carried away with predictive cases, so try to limit your study to four or five main predictions, and after an economic analysis, expand on the most likely case. A good rule of thumb is to consider a predictive case accurate for twice as long as the history match as long as the predictive mechanisms are not different from those in the history match. Considering economic evaluations based on a present value concept, changes in oil production after ten years have little effect on the present worth.

One very straightforward method of predicting is to use a fixed production schedule by specifying oil producing rates and injection rates; however as gas and water production increase, this usually requires a large number of rate adjustments. Use of GOR and WOR limits in conjunction with bottom-hole pressure controls can alleviate some of the manual intervention. By controlling production using bottom-hole pressure and productivity indexes, predictions may require very little manual intervention. Another useful item is to specify a liquid producer for wells on artificial lift; as the WOR (or WCUT) changes, the total fluid volume remains constant. For reinjection schemes, fractional injectors (i.e., wells that inject a fraction of the produced water or gas) will avoid frequent manual rate adjustments.

Late in predictive cases, particularly when saturation and pressure changes are minimal, large timesteps may be employed even though this introduces greater error; usually, the savings in computer time outweighs the error introduced in present worth calculations as previously explained. More information on predicting well performance may be found in Chap. 9 (pp. 99–110) of M-13.

Problems — Chapter 8

1. *History matching.* a. A reservoir with an active water drive is being modeled. Average model pressure is currently below the actual field value. What do you adjust first in order to attempt to match history?

b. After matching average pressure, your simulated water production is greater than that in the field. What do you adjust now?

c. After matching water production data, your bottom-hole pressures are still low. Core analysis and logs from 5 wells indicate consistent rock properties. What do you adjust to correct the bottom-hole pressures?

2. *History Matching.* a. A reservoir is being modeled and the simulator gas production is lower than actual field values; the average pressure in the model is higher than observed data, and the reservoir is well below bubble point. What do you adjust to match history?

b. In question 2a, if the reservoir is undersaturated, what would you adjust to match history?

c. In question 2a, if the average model pressure is lower than the field data, what would you adjust?

Chapter 9

OTHER TYPES OF MODELS

Strictly speaking, compositional simulators should be used for many enhanced recovery studies; however the complexity of these models requires a simulation specialist and as such removes them from the "routine-use" list. To allow complex recovery mechanisms to be studied, especially for screening purposes, variations of the black oil simulator have evolved for specific recovery processes. These usually consist of oil, gas, water, and a fourth (or more) component (e.g. miscible injectant or polymer) and require only a slight amount of additional data over a black oil model.

Polymer models are used to simulate polymer solution injection in a heterogeneous reservoir system. Polymer behavior is modeled by treating the polymer solution as a fourth phase. Various concentrations of the polymer may be injected as in a tapered slug process. Slug dispersion is modeled through use of a **mixing parameter**, Ω, which varies from zero (no mixing) to one (complete mixing). The mixing parameter also accounts for gridding effects and while it may be estimated from sensitivity testing, it actually becomes another history match parameter and is best approximated from pilot studies. Polymer adsorption in micrograms polymer per gram of formation and the formation grain density in gm/cc are required to calculate the amount of polymer lost to the formation; an average grain density for sandstone is 2.87. The residual resistance effect

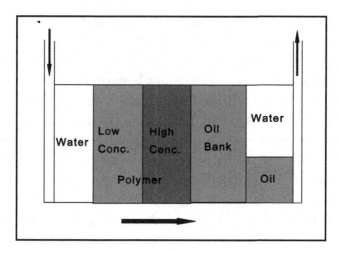

Schematic polymer recovery

(due primarily to permeability reduction) is expressed as a ratio of water mobility prior to polymer contact to water mobility after polymer contact.

Near-well injectivity effects are modeled for various concentrations using a logarithmic relationship. All of the above-mentioned data (except the mixing parameter) are usually obtained from a

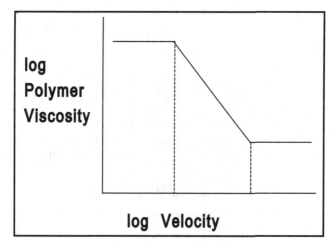

Polymer viscosity–velocity plot

laboratory analysis or existing literature for the polymer under consideration. Additionally, the polymer concentration must be specified for the injection well. This model will require approximately half a dozen additional input lines of data as compared to a black oil model. Additional information on both polymer and surfactant flooding is in Sec. 11.3, p. 116, M-13.

Miscible models are used to simulate miscible flow performance by injection of a gas above its miscibility pressure. This behavior is modeled by treating the injected fluid as a fourth phase below miscibility pressure and as part of the oil phase above miscible conditions. Additional data requirements include the miscibility pressure and a minimum saturation for miscibility to exist in a cell. As with the polymer model, an empirical mixing parameter, Ω, is needed, and the same discussion is applicable, except that miscible models are much more sensitive to slight variations in the mixing parameter (than are polymer models). A density of the injected fluid (usually at standard conditions), and the volume factors and viscosities (as functions of pressure) are also required. Much of this data is available in the literature or from lab analysis. Mobility reduction due to

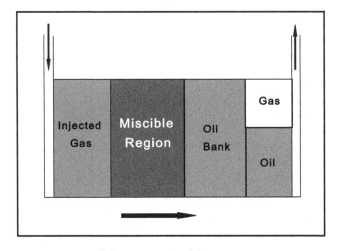

Schematic miscible recovery

viscous mixing is calculated. Approximately a dozen additional lines of input data are needed. Miscible models are much more sensitive than black oil models due to loss or attaining of miscibility, and due to alternate injection of gas and water. Again, this type of model is intended for use as a screening tool. A more detailed explanation of miscible displacement may be found in Sec. 11.2, p. 115, M-13.

9.1. Radial Simulators

Radial simulators, sometimes known as coning models, or simply R-Z simulators are one-well models that are used primarily to study well tests, coning phenomena, induced fracture effects, well completions and other individual well factors. This type of model must be fully implicit to account for the rapid saturation changes that will occur in and around the wellbore. While the lengths of the radial cells may be specified, it is usually much easier to specify the wellbore and the drainage radius, and allow the model to determine intermediate values based on logarithmic spacing. Usually ten cells in the radial direction are

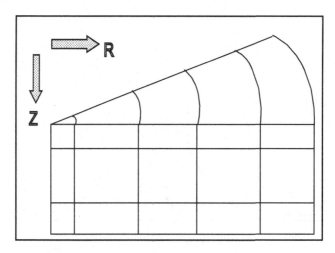

Radial model

sufficient and an upper limit of twenty cells should be considered. Vertical layering is determined by stratification, fluid contacts and sufficient definition in zones of interest. While wells may be positioned at cells other than the center, such wells are merely injection and withdrawal points, and will not exhibit any individual well characteristics; similarly, water injection wells may be used in the last ring of cells to simulate an aquifer; however, the use of a pore volume multiplier is usually preferred.

Coning is a phenomenon where water is drawn up into the wellbore due to the production rate. Since water is more dense than oil, there are gravity forces opposing the growth of the water cone. For a given oil production rate there is a particular height above the water–oil boundary at which the cone's apex will come to rest in equilibrium. At this point pressure and gravity forces are balanced. There exists a critical production rate above which no stable cone can be formed. The cone will continue to rise until it has reached the well and water will be produced along with the oil. In exactly the same manner, a water cone can be formed in a water–gas system. In a gas–oil system,

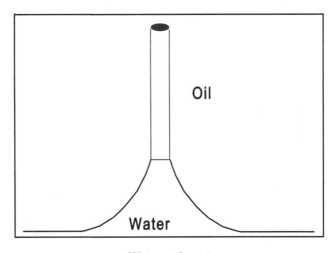

Water–oil coning

a gas cone can be formed. This phenomena differs from water coning only in two respects: the cone is obviously composed of gas and the cone is upside down. A knowledge of the critical coning rate for a particular well would be very useful information. It would tell us whether or not coning could be expected to be a problem. It might help us decide at what depth the well should be completed, or whether or not we should take advantage of any shale barriers present. The results might indicate that the critical rate is very low and to produce below it would not be economically feasible. If the critical rate is in fact too low to be practical, we may choose to produce at a higher rate. In this case we would be interested in how long it will be before water or gas will break through and be produced, and what amount will be produced. A layered radial model will exhibit these effects; however, if we wish to recomplete the well to avoid further coning problems, then flow reversal may occur and hysteresis effects must be included in the simulator. The same is true for rate variations to cause cone reductions.

Well tests are easily modeled with a radial simulator. The only difference between modeling a well test and other types of simulations is that exceptionally small timesteps will be used to match the test data. Usually, modeling a well test is not sufficient to yield a unique reservoir description.

Induced fracture modeling can be performed using the angular direction, Θ, in a radial model. A very slight angle is used for the fracture, and it is given a large permeability in both the radial and angular directions. Modifications may be required to outer transmissibilities because even very small angles generate a large pore volume over a distance. Radial and vertical transmissibilities are sealed at the edges (similar to a cartesian system), but the last angular transmissibility must be the first angular transmissibility when a full circle is involved.

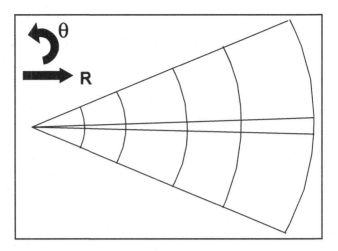

Radial fracture model

9.2. Dual Porosity Simulators

Dual porosity models are used to simulate naturally fractured reservoirs. They allow two sets of porosity and permeability values: one for the rock matrix and one for the fracture system.

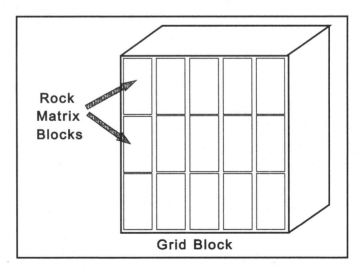

Dual porosity model

If production is mainly from the fractures, two porosities are needed, but only one permeability (the rock matrix feeds the fractures and no flow occurs through the rock matrix). The volume within the fractures is usually small resulting in the rock matrix as the source of oil and gas. This type of model should be fully implicit. Both cartesian and radial geometry systems may be employed. The fractures will have high permeabilities and low porosities while the rock matrix will have low (or no) permeabilities and high porosities. Fracture properties are averaged using a volume containing both the fracture and rock matrix, thus resulting in low fracture porosity values. A shape factor, σ, is used to determine the conductivity between the rock matrix and the surrounding fractures. For rectangularly shaped matrix blocks of dimensions L_x, L_y, and L_z, the shape factor may be calculated using

$$\sigma = 4 \left(\frac{1}{L_x^2} + \frac{1}{L_y^2} + \frac{1}{L_z^2} \right).$$

Often, the shape factor becomes an additional history match parameter. Fracture relative permeabilities use a vertical equilibrium concept and are usually straight lines; fracture capillary pressures are normally zero.

Additional information concerning dual porosity models may be found on p. 25, M-13.

Chapter 10

ODDS AND ENDS

Restart capability allows a model to be started again at any point in time (previously simulated). It consists of storing all of the necessary data at various points in time. This feature has two main uses: for long history matches, it is not necessary to complete the history to check intermediate results, and all predictive cases may be started from the same history match.

Preprocessors are used for gridding and interpolating data. **Postprocessors** such as **plotting routines, summary routines** and **contouring programs** summarize data so that it may be rapidly interpreted. For large studies, use of graphical results aids in history matching and allows a clear comparison of several predictive cases. Use of these routines does not eliminate normal model monitoring requirements.

Currently, a wide variety of **computers** are available for reservoir simulation. Mainframe computers have tremendous storage capacities but have been replaced by workstation systems in most companies. Additional solution speed is achieved when using parallel processing. Workstation systems (implying intermediate size computers) fall between PC and mainframe capabilities, and are one of the most popular systems based on cost and multitasking capability. PCs' are slower and may be cell-limited due to memory limitations although small grid systems (up to 1000 cells) will fit in active memory and larger grids may be used with expanded memory options; suffice to say, large grids may run slowly on PCs due to either processor speed, RAM limitations and data transfer speed.

The following items should be considered in **selecting an appropriate simulator:**

1. Reservoir type (i.e., homogeneous, layered, etc).
2. Data availability.
3. Mechanisms affecting performance (i.e., water underrunning).
4. Answers required.
5. Level of precision in answers required.
6. Amount of engineering time available.
7. Amount of computer time/expense.

A **checklist of items to review when initiating a simulation study** would include the following:

- Grid system selection
- Fluid property selection
- IMPES or Fully implicit
- Equilibrium initialization
- Saturation-dependent data
    - Capillary pressure — back calculate from known saturations
    - Relative permeability — lab data or correlations
    - Irreducible water saturation — $1/2 \to 1\%$ above S_{wc}
    - 3-phase k_{ro} — Stone statistical method (Stone 1)
    - Use two-point upstream weighting
- PVT data
    - Smooth data
    - Compare with known correlations for consistency
    - If equally-spaced, watch extrapolation (particularly on B_o)
- Depths — eliminate micro highs (or micro lows)
- Pore volumes — compare adjacent cells (factor of $2 \to 5$)

Difficulty in solving may be related to several possibilities, but a few of the initial items to be checked are

1. Maximum saturation and pressure changes over the timestep.
2. Cell in which maximum changes occur (is it a well cell?).
3. Material balances.
4. Timestep size.
5. Well rates and location.
6. Pore volume throughput and its associated well.
7. Pressure change over the last iteration.

If the maximum changes are not occurring in a well cell, either a very small pore volume cell is located next to a large pore volume cell, or a very rapidly moving front exists (double check relative permeability data in this region). If the problem occurs when going through a saturation pressure, check the solution gas, formation volume factors, and oil compressibilities for correctness and smoothness; it may be necessary to compare the first and second derivatives to determine oscillations that may not appear graphically.

Expect to use small timesteps when dissolving the last bubble of gas in a cell due to repressuring; in an IMPES model in particular, this situation is subject to overshoot (on gas saturation).

Finally, a brief review of the advantages and disadvantages of reservoir simulation would include the following items.

10.1. Advantages of Reservoir Simulation

1. The analytical limitations of simpler methods are overcome.
2. Data variation within a reservoir can be applied; homogeneity is not a requirement.
3. The effect of uncertainty in the reservoir description can be analyzed with sensitivity testing (i.e., what if the

permeability is doubled?); reviewing the available data often leads to "weak links" that were previously unknown.

4. After matching history, many different methods of operating the reservoir in the future can be investigated and an optimum plan of reservoir management can be formulated.

5. Continual performance monitoring is available.

6. The computational burden is reduced for the engineer and additional time is available for analyzing results.

7. A common tool is employed in arbitration and unitization decisions.

10.2. Disadvantages of Reservoir Simulation

1. Modeling requires a significant amount of reasonable data.

2. Modeling requires a significant amount of knowledgeable manpower and time.

3. Simulation has limitations that a casual user/observer may not fully comprehend (i.e., the effect of cell size).

4. Software/hardware costs are greater than analytical methods.

5. Results are not unique (i.e., the same answer can be obtained by varying several different parameters).

APPENDIX A

A.1. Fluid and Formation Correlations

Most simulation studies will have a majority of the fluid (PVT) and formation properties available from a laboratory analysis; however, some properties (e.g. water compressibility) are seldom measured in the laboratory. For some preliminary studies, it may be necessary to estimate a majority of the properties. A number of fluid property correlations have been proposed, among them:

Standing	(California Crudes)
Vazquez & Beggs	(Worldwide Crudes)
Glaso	(North Sea Crudes)
Al-Marhoun	(Middle East Crudes)
Petrosky & Farshad	(Gulf of Mexico Crudes)
Dokla & Osman	(U.A.E. Crudes)
Farshad	(Columbian Crudes)
Obomanu & Okpobiri	(Nigerian Crudes)
Kartoatmodjo & Schmidt	(Worldwide Crudes)

The following correlations may prove useful.

Solution gas: One correlation which has yielded fairly uniform results is that of Standing,

$$R_s = \gamma_g[(P + 25.48)/(18.2^*10^4)]^{1.205}$$

and

$$A = 0.00091\,T - 0.0125\,\text{API},$$

where

$$R_s = \text{solution gas, SCF/STB}$$
$$\gamma_g = \text{gas gravity}$$
$$P = \text{reservoir pressure, psia}$$
$$T = \text{reservoir temperature, } °\text{F}$$
$$\text{API} = \text{oil gravity, } °\text{API.}$$

Vazquez and Beggs present a correlation dependent on the range of the oil gravity,

$$R_s = C1\gamma_g P^{C2} e^{AA}$$

and

$$AA = C3[\text{API}/(T + 460)].$$

The constants $C1, C2,$ and $C3$ are API-dependent

	API $\leq$ 30	API $>$ 30
$C1$	0.0362	0.0178
$C2$	1.0937	1.187
$C3$	25.724	23.931 .

Glaso's correlation uses

$$BB = 2.8869 - (14.1811 - 3.3093 \log(P))^{0.5}$$

to calculate

$$R_s = \gamma_g((\text{API}^{0.989}/T^{0.172}) \, 10^{BB})^{1.2255}.$$

Al-Marhoun requires the oil specific gravity,

$$\gamma_o = 141.5/(131.5 + \text{API})$$

to calculate the dissolved gas

$$R_s = \left\{ 185.843208 \gamma_g^{1.87784} \gamma_o^{-3.1437} (T + 460)^{-1.32657} P \right\}^{1.398441}.$$

And Petrosky & Farshad have developed the following:

$$R_s = \left\{ [(P/112.727) + 12.34] \gamma_g^{0.8439} 10^X \right\}^{1.73184},$$

where

$$X = 0.0007916\,\text{API}^{1.541} - 0.00004561\,T^{1.3911}.$$

Dokla & Osman use

$$R_s = \left[1.19562 * 10^{-4} P \gamma_g^{1.01049} \gamma_o^{-0.107941} (T + 460)^{0.952584} \right]^{1.3811258}.$$

While Farshad recommends the following for Columbian crudes,

$$R_s = 0.01936 P^{1.1574} \gamma_g^{0.73495} 10^{0.000337\,T + 0.01771\,\text{API}}.$$

Obomanu & Okpobiri use

$$R_s = 5.615 \left[0.0308(6.894\ P)^{0.927} \gamma_g^{2.15} \text{API}^{1.27} \right] / (10^{0.881} T^{0.497}).$$

And finally, Kartoatmodjo and Schmidt offer a selection based on API gravity for API > 30,

$$R_s = 0.0315 \gamma_g^{0.7567} P^{1.0937} 10^{11.2895[\text{API}/(T+460)]}$$

for API $\leq$ 30,

$$R_s = 0.05958 \gamma_g^{0.7972} P^{1.0014} 10^{13.1405[\text{API}/(T+460)]}.$$

Oil formation volume factor: A reasonable estimate of B_o below (and including) the bubble point is Standing's correlation

$$B_o = 0.9759 + 0.00012\ X^{1.2}$$

and

$$X = R_s\ (\gamma_g/\gamma_o)^{0.5} + 1.25\ T,$$

where

$$B_o = \text{oil formation volume factor RVB/STB}$$
$$\text{(below or at bubble point)}$$
$$R_s = \text{solution gas, SCF/STB}$$
$$\gamma_g = \text{gas gravity}$$
$$\gamma_o = \text{oil specific gravity}$$
$$T = \text{reservoir temperature, }^\circ\text{F}$$
$$\text{API} = \text{oil gravity, }^\circ\text{API.}$$

Vazquez and Beggs again differentiate using two ranges of oils

$$B_o = 1 + C1\,R_s + (T - 60)(\text{API}/\gamma_g)(C2 + C3\,R_s),$$

where the constants $C1, C2$, and $C3$ are API-dependent

	API $\leq$ 30	API $>$ 30
$C1$	0.0004677	0.000467
$C2$	$1.751 * 10^{-5}$	$1.100 * 10^{-5}$
$C3$	$1.811 * 10^{-8}$	$1.337 * 10^{-9}$.

Glaso's correlation uses

$$G = R_s(\gamma_g/\gamma_o)^{0.526} + 0.968\,T$$
$$H = -6.58511 + 2.91329\log(G) - 0.27683\log(G)^2$$
$$B_o = 1 + 10^H.$$

Al-Marhoun employs the following equations

$$F = R_s^{0.74239}\gamma_g^{0.323294}\gamma_o^{-1.20204}$$
$$B_o = 0.497069 + 8.62963 * 10^{-4}(T + 460) + 0.82594 * 10^{-3}F$$
$$+ 3.18099 * 10^{-6}F^2.$$

And for Gulf of Mexico crude oils, Petrosky & Farshad present

$$A = \left\{ R_s^{0.3738} \left(\gamma_g^{0.2914} / \gamma_o^{0.6265} \right) + 0.24626T^{0.5371} \right\}^{3.0936}$$
$$B_o = 1.0113 + 7.2046 * 10^{-5} A.$$

Dokla & Osman present

$$B_o = 0.0431935 + 0.00156667(T + 460) + 0.00139775F$$
$$+ 3.80525 * 10^{-6} F^2,$$

where

$$F = R_s^{0.773572} \gamma_g^{0.40402} \gamma_o^{-0.882605}.$$

For Columbian crudes, Farshad recommends

$$B_o = 1.0 + 10^{AA}$$

with

$$A = R_s^{0.5956} \gamma_g^{0.2369} \gamma_o^{-0.13282} + 0.0976 \, T$$

and

$$AA = -2.6541 + 0.5516 \, \log(A) + 0.331 \, \log(A^2).$$

Obomanu & Okpobiri use the following for Nigerian crude oils, for API > 30,

$$B_o = 0.3321 + 7.88374 * 10^{-4} R_s / 5.615$$
$$+ 2.335 * 10^{-3} R_s \gamma_g / (5.615 \gamma_o) + 2.0855 * 10^{-3} [5(T+460)/9]$$

for API ≤ 30,

$$B_o = 1.0232 + 1.065 * 10^{-4} \{ R_s[(\gamma_g/\gamma_o) + T]/5.615 \}^{0.79}.$$

Kartoatmodjo & Schmidt employ

$$B_o = 0.98496 + F^{1.5}/1000,$$

where

$$F = R_s^{0.755} \gamma_g^{0.25} \gamma_o^{-1.5} + T.$$

Above the bubble point, all of the previous correlations require a correction to the oil formation volume factor using oil compressibility

$$B_o = B_{oBP} \, e^{-c_o*(P-P_{BP})}$$

which is often shown using a power series expansion as

$$B_o = B_{oBP} \, [1 - c_o \, (P - P_{BP})],$$

where

$$B_o = \text{oil formation volume factor RVB/STB}$$
$$\text{(above bubble point)}$$
$$B_{oBP} = \text{oil formation volume factor RVB/STB}$$
$$\text{(at bubble point)}$$
$$c_o = \text{undersaturated oil compressibility, /psi}$$
$$P = \text{reservoir pressure, psi}$$
$$P_{BP} = \text{bubble point pressure, psi.}$$

Gas formation volume factor: The gas formation volume factor is readily calculated from

$$B_g = 5.035 \, z \, (T + 460)/P,$$

where

$$B_g = \text{gas formation volume factor RVB/MCF}$$
$$z = \text{gas deviation factor}$$
$$T = \text{reservoir temperature, }^\circ\text{F}$$
$$P = \text{reservoir pressure, psia.}$$

Values of the gas deviation factor (z-factor) may be obtained from laboratory analysis of gas samples or correlations. All reasonable correlations are iterative and probably the most popular correlation for gas deviation factors is that of Yarborough and Hall (Oil & Gas Journal, June 18, 1973; Feb. 18, 1974).

This technique may be readily solved using a Newton–Raphson iteration scheme. The technique should not be used at reduced temperatures of less than one or at reduced pressures above 20.5. The method employs the use of a hard-sphere equation of the form

$$f(y) = A\,{}_pP_r + \frac{y + y^2 + y^3 - y^4}{(1-y)^3} - B\,y^2 + C\,y^D$$

and the derivative is

$$\frac{\partial f(y)}{\partial y} = \frac{1 + 4y + 4\,y^2 - 4\,y^3 + y^4}{(1-y)^4} - E\,y + CD\,y^F.$$

Using the Newton–Raphson scheme,

$$y_{\text{new}} = y - \frac{f(y)}{\frac{\partial f(y)}{\partial y}},$$

where

$$t = 1/{}_pT_r$$
$$A = -0.06125te^{A1}$$
$$A1 = -1.2(1-t)^2$$
$$B = 14.7t - 9.76t^2 + 4.58t^3$$
$$C = 90.7t - 242.2t^2 + 42.2t^3$$
$$D = 2.18 + 2.82t$$
$$E = 29.52t - 19.52t^2 + 9.16t^3$$
$$F = 1.18 + 2.82t.$$

The gas deviation factor is calculated using

$$z = \frac{-A\,{}_pP_r}{y}$$

when y and y_{new} are within 10^{-5}.

Note that

$$_pT_r = \text{pseudoreduced temperature} = {_pT_c}/(T + 460)$$
$$T = \text{reservoir temperature,}\,^\circ\text{F}$$
$$\gamma_g = \text{gas gravity}$$
$$P = \text{reservoir pressure, psia}$$
$$_pP_r = \text{pseudoreduced pressure} = {_pP_c}/P$$
$$z = \text{gas deviation factor}$$

and the pseudocritical properties may be obtained for a dry gas using

$$_pT_c = 168 + 325\ \gamma_g - 12.5\ \gamma_g^2$$
$$_pP_c = 677 + 15\ \gamma_g - 37.5\ \gamma_g^2,$$

and for a wet gas

$$_pT_c = 187 + 330\ \gamma_g - 71.5\ \gamma_g^2$$
$$_pP_c = 706 - 51.7\ \gamma_g - 11.1\ \gamma_g^2.$$

Water formation volume factor: The water formation volume factor may be calculated at reservoir pressures using

$$B_w = B_{wb}\ e^{-c_w\ (P - P_b)}$$

or employing a power series expansion,

$$B_w = B_{wb}\ [1 - c_w\ (P - P_b)],$$

where

$$B_w = \text{water formation volume factor RVB/STB}$$
$$B_{wb} = \text{water formation volume factor at } P_b, \text{RVB/STB}$$
$$c_w = \text{water compressibility, /psi}$$
$$P = \text{reservoir pressure, psia}$$
$$P_b = \text{base pressure, psia.}$$

B_w may be estimated as a function of pressure from

$$B_w = -1.485 * 10^{-6} P + 0.952 + 10^A$$

and

$$A = 0.001996 \, (T - 100) - 1.267606,$$

where

B_w = water formation volume factor, RVB/STB

P = reservoir pressure, psia

T = reservoir temperature, °F.

Oil viscosity: Dead-oil viscosity (with no solution gas, or at a base pressure) may be estimated from data by Chew and Connally,

$$\mu_{DO} = \{0.32 + [(1.8 * 10^7)/(\text{API}^{4.53})]\} * [360/(T + 200)]^D$$

and

$$D = 10^{[0.43 + (8.33/\text{API})]},$$

where

μ_{DO} = dead-oil viscosity, cp

API = oil gravity, °API

T = reservoir temperature, °F.

Live-oil viscosity (including solution gas) may be estimated at pressures up to (and including) the bubble point by data from Beal

$$\mu_o = Q * \mu_{DO}{}^R$$

and

$$Q = 10^M$$
$$M = 2.2 * 10^{-7} \, R_s^2 - 7.4 * 10^{-4} \, R_s$$
$$R = (0.68/10^A) + (0.25/10^B) + (0.062/10^C)$$
$$A = 8.62 * 10^{-5} \, R_s$$
$$B = 1.10 * 10^{-3} \, R_s$$
$$C = 3.74 * 10^{-3} \, R_s,$$

where

$$\mu_o = \text{oil viscosity, cp (at or below b.p.)}$$
$$\mu_{DO} = \text{dead oil viscosity, cp}$$
$$R_s = \text{solution gas, SCF/STB.}$$

Above the bubble point, oil viscosity may be determined from

$$\mu_o = \mu_{oBP} + Y \left(P - P_{BP} \right)$$

and

$$Y = 0.001 e^X$$
$$X = -2.68 + 0.98 \ \ln(\mu_{oBP}) + 0.091 \ [\ln(\mu_{oBP})]^2,$$

where

$$\mu_o = \text{oil viscosity, cp (above b.p.)}$$
$$\mu_{oBP} = \text{oil viscosity at b.p., cp}$$
$$P = \text{reservoir pressure, psia}$$
$$P_{BP} = \text{bubble point pressure, psia.}$$

Gas viscosity: When unavailable as laboratory data, gas viscosities may be estimated from the following equations of Lee *et al.* (JPT, Aug. 1966):

$$\mu_g = 10^{-4} Y1 \ e^H$$
$$\text{MWG} = 28.9 \ \gamma_g$$
$$Y1 = [(9.4 + 0.02 \ \text{MWG}) * (T + 460)^{1.5}]/(209$$
$$+ 19 \ \text{MWG} + T + 460)$$
$$Y2 = 3.5 + 986/(T + 460) + 0.01 \ \text{MWG}$$
$$Y3 = 2.4 - 0.2 \ Y2$$
$$Y4 = 0.00752 \ \text{MWG}/B_g$$

and

$$H = Y2 \, Y4^{Y3},$$

where

MWG = gas molecular weight

T = reservoir temperature, °F

B_g = gas formation volume factor, RVB/MCF

μ_g = gas viscosity, cp.

Water viscosity: A correlation that might prove useful is

$$\mu_w = 4.33 - 0.07T + 4.73 * 10^{-4}T^2$$
$$- 1.415 * 10^{-6}T^3 + 1.56 * 10^{-9}T^4,$$

where

μ_w = water viscosity, cp

T = reservoir temperature, °F.

Oil compressibility: A correlation by Vazquez and Beggs (JPT, June 1980) that has proven satisfactory is

$$c_o = (-1433 + 5000\,R_{sBP} + 17.2\,T - 1180\,\gamma_g$$
$$+ 12.61\,\text{API})/(10^5 P),$$

where

c_o = undersaturated oil compressibility, /psi

R_{sBP} = dissolved gas at bubble point, MCF/STB

T = temperature, °F

γ_g = gas gravity

API = oil gravity, °API

P = pressure, psi.

Beggs defines the gas gravity at specific separator conditions, but for this correlation, it is seldom critical.

Petrosky offers the following correlation for undersaturated oil compressibility

$$c_o = 1.705 * 10^{-7} R_s^{0.69357} \gamma_g^{0.1885} \text{API}^{0.3272} T^{0.6729} P^{-0.5906}.$$

Water compressibility: c_w may be estimated from the equation

$$c_w = (A + B \, T + C \, T^2) * 10^{-6}$$

and

$$A = 3.8546 - 1.34 * 10^{-4} \, P$$
$$B = -1.052 * 10^{-2} + 4.77 * 10^{-7} \, P$$
$$C = 3.9267 * 10^{-5} - 8.8 * 10^{-10} \, P,$$

where

$$c_w = \text{water compressibility, /psi}$$
$$T = \text{temperature, } °\text{F}$$
$$P = \text{pressure, psia.}$$

Formation compressibility: No one seems satisfied with the current formation compressibility correlations; however, here are some for your selection. A general correlation is that of Hall,

$$c_f = [13.392/(\phi^{0.438})] * 10^{-6},$$

where

$$c_f = \text{formation compressibility, /psi}$$
$$\phi = \text{porosity, \%.}$$

Another correlation is that of Van der Knaap (using the same units as Hall),

$$c_f = [100/\phi^{1.2}] * 10^{-6}.$$

Newman presents four different rock types (using the same units as Hall), for consolidated sandstones,

$$c_f = [40/\phi^{0.75}] * 10^{-6}$$

for friable sandstones,

$$c_f = (10^{-0.0073\phi + 1.415}) * 10^{-6}$$

for unconsolidated sandstones,

$$c_f = (10^{0.03925\phi+0.3405}) * 10^{-6}$$

and for carbonates,

$$c_f = [40/\phi^{0.67}] * 10^{-6}.$$

Yale *et al.* offer a general equation of the form

$$c_f = a(\sigma - b)^c + d,$$

where

$$\sigma = (k1 + k2 + k3)p$$

$$p = \text{reservoir pressure, psia}$$

and the constants are defined for various formations as

	a	b	c	d	$k1$	$k2$	$k3$
Consolidated sandstones	$-2.399 * 10^{-5}$	300	0.0623	$4.308 * 10^{-5}$	0.85	0.80	0.45
Friable sandstones	$-1.054 * 10^{-4}$	500	-0.225	$1.1103 * 10^{-5}$	0.90	0.90	0.60
Unconsoli-dated sands	$-2.805 * 10^{-5}$	300	0.1395	$1.183 * 10^{-4}$	0.95	0.95	0.75
Carbonates	$-2.399 * 10^{-5}$	300	0.0623	$4.308 * 10^{-5}$	0.85	0.85	0.55.

Relative permeability: The following correlations are by Honarpour, Koederitz & Harvey (SPE Trans., 1982).

Sandstones and Conglomerates:

Water-wet system:

$$k_{rw} = 0.035388\frac{S_w - S_{wc}}{1 - S_{wc} - S_{orw}} - 0.010874\left(\frac{S_w - S_{orw}}{1 - S_{wc} - S_{orw}}\right)^{2.9}$$
$$+ 0.56556 (S_w - S_{wc}) S_w^{3.6}.$$

Intermediate and oil-wet systems:

$$k_{rw} = 1.5814\left(\frac{S_w - S_{wc}}{1 - S_{wc}}\right)^{1.91} - 0.58617\frac{S_w - S_{orw}}{1 - S_{wc} - S_{orw}} (S_w - S_{wc})$$
$$- 1.2484\phi(S_w - S_{wc})(1 - S_{wc}).$$

Any wettability system:

$$k_{row} = 0.76067 \left(\frac{\frac{S_o}{1-S_{wc}} - S_{orw}}{1 - S_{orw}} \right)^{1.8} \left(\frac{S_o - S_{orw}}{1 - S_{wc} - S_{orw}} \right)^2$$

$$+ 2.6318\phi \left(1 - S_{orw} \right) \left(S_o - S_{orw} \right),$$

$$k_{rog} = 0.98372 \left(\frac{S_o}{1 - S_{wc}} \right)^4 \left(\frac{S_o - S_{org}}{1 - S_{wc} - S_{org}} \right)^2,$$

$$k_{rg} = \left[1.1072 \left(\frac{S_g - S_{gc}}{1 - S_{wc}} \right)^2 + 2.7794 \frac{S_{org} \left(S_g - S_{gc} \right)}{1 - S_{wc}} \right] k_{rg@\,S_{org}}.$$

Limestones and Dolomites:

Water-wet system:

$$k_{rw} = 0.0020525 \frac{S_w - S_{wc}}{\phi^{2.15}} - 0.051371(S_w - S_{wc}) \left(\frac{1}{k} \right)^{0.43}.$$

Intermediate and oil-wet systems:

$$k_{rw} = 0.29986 \frac{S_w - S_{wc}}{1 - S_{wc}} - 0.32797 \left(\frac{S_w - S_{orw}}{1 - S_{wc} - S_{orw}} \right)^2 (S_w - S_{wc})$$

$$+ 0.413259 \left(\frac{S_w - S_{wc}}{1 - S_{wc} - S_{orw}} \right)^4.$$

Any wettability system:

$$k_{row} = 1.2624 \frac{S_o - S_{orw}}{1 - S_{orw}} \left(\frac{S_o - S_{orw}}{1 - S_{wc} - S_{orw}} \right)^2,$$

$$k_{rog} = 0.93752 \left(\frac{S_o}{1 - S_{wc}} \right)^4 \left(\frac{S_o - S_{org}}{1 - S_{wc} - S_{org}} \right)^2,$$

$$k_{rg} = 1.8655 \frac{(S_g)(S_g - S_{gc})}{1 - S_{wc}} k_{rg@\,S_{org}} + 8.0053 \frac{(S_g - S_{gc})(S_{org})^2}{1 - S_{wc}}$$

$$- 0.025890(S_g - S_{gc}) \left(\frac{1 - S_{wc} - S_{org} - S_{gc}}{1 - S_{wc}} \right)^2$$

$$\times \left(\frac{S_{org} + S_{gc}}{1 - S_{wc}} \right)^2 \sqrt{\frac{k}{\phi}},$$

where

k_{rw} = water relative permeability, fraction

k_{row} = oil relative permeability with respect to water, fraction

k_{rg} = gas relative permeability, fraction

k_{rog} = oil relative permeability with respect to gas, fraction

S_w = water saturation, fraction

S_{wc} = critical (connate) water saturation, fraction

S_o = oil saturation, fraction

S_{orw} = residual oil saturation with respect to water, fraction

S_{org} = residual oil saturation with respect to gas, fraction

S_g = gas saturation, fraction

S_{gc} = critical gas saturation, fraction

ϕ = porosity, fraction

$k_{rg@Sorg}$ = gas relative permeability at residual oil, fraction

k = absolute permeability, md.

A more recent and larger set of relative permeability correlations may be found in SPE 65631, "Two-Phase Relative Permeability Prediction Using a Linear Regression Model," by Mohamad Ibrahim and Koederitz.

Solutions to Problems

Solutions — Chapter 1

1. *Solution*: Employ a material balance to check the pressures and required injection prior to a simulation study; various steady state injection calculations would also be appropriate. Incidentally, the grid shown below was initially proposed but would yield extremely optimistic results.

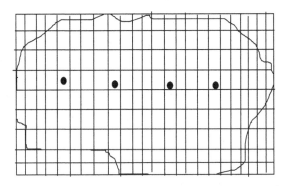

9 x 24 = 216 cells

Actual areal grid

An areal grid was employed for the actual study and indicated that one of the center wells should be converted to injection (note that there are only two locational injection schemes: inner and outer). Two years of production history existed at the initiation of this study and the history match indicated a

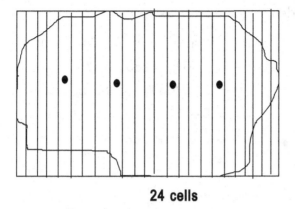

24 cells

Incorrect linear grid

fairly uniform drawdown throughout the reservoir; both analytical solutions and the simulation indicated an additional two years prior to approaching the bubble point pressure. The wells were pressure tested at three years to validate the simulation study and the west half of the field indicated a higher pressure than predicted while the eastern half was low and rapidly approaching the bubble point. Interference testing then indicated a lack of communication between the two center wells and the possible presence of a fault. After including the fault in the simulator, the east half matched known historical performance. The western half was low in pressure and the presence of an active aquifer was suspected, resulting in downsizing the injection facilities as only the east end required pressure support.

2. *Solution*:

$$r = x/\cos\theta$$
$$= 5280/\cos 6° = 5280/0.9945 = 5309'$$
$$\text{error} = 1 - \cos\theta = 1 - 0.9945 = 0.0055 = 0.55\%.$$

Actually, defining the error as

$$\begin{aligned}
\text{error} &= (\text{true} - \text{approximate})/\text{true} \\
&= (5309 - 5280)/5309 \\
&= 5309/5309 - 5280/5309 \\
&= 1 - x/r \\
&= 1 - \cos\theta.
\end{aligned}$$

3. *Solution*:

$$\begin{aligned}
\phi_2 &= \phi_1 e^{c_f(p_2 - p_1)} \\
&= (0.20)e^{(0.0000036)(1000 - 5000)} \\
&= 0.1971
\end{aligned}$$

or

$$\begin{aligned}
\phi_2 &= \phi_1[1 + c_f(p_2 - p_1)] \\
&= (0.20)[1 + (3.6 \times 10^{-6})(1000 - 5000)] \\
&= 0.1971.
\end{aligned}$$

NOTE: For an extremely compressible formation having a compressibility of 100 microsips,

$$\phi_2 = 0.134 \text{ using } \phi_2 = \phi_1 e^{c_f(p_2 - p_1)}$$

however,

$$\phi_2 = 0.120 \text{ using } \phi_2 = \phi_1[1 + c_f(p_2 - p_1)].$$

$$\begin{aligned}
\phi_2 &= \phi_1 e^{c_f(p_1 - p_2)} \\
&= (.20)e^{(0.0000036)(100 - 5000)} \\
&= 0.1965.
\end{aligned}$$

4. *Solution*:

$$k_{\text{horizontal}} = [(140)(7) + (20)(15) + (70)(4)]/(7 + 15 + 4) = 60\,\text{md}$$

$$k_{\text{vertical}} = (7 + 15 + 4)/[(7/140) + (15/20) + (4/70)] = 30.3\,\text{md}$$

$$k_{\text{geometric}} = e^{[7\ln(140) + 15\ln(20) + 4\ln(70)]/(7 + 15 + 4)} = 40.95\,\text{md}.$$

If the evenly-spaced geometric average is calculated (in error),

$$k_{\text{geo-even}} = [(140)(20)(70)]^{1/3} = 58.09 \, \text{md}.$$

5. *Solution*:

$$(5280)(\tan 8°) = 742 \, \text{ft/mile}$$
$$(3)(742) = 2226 \, \text{ft}.$$

And estimating the additional pressure using a brine hydrostatic gradient of 0.466 psi/ft results in

$$(2226)(0.466) = 1037 \, \text{psi}.$$

6. *Solution*:

$$P = P_{\text{datum}} + \rho \Delta D / 144$$
$$= 2200 + (52.1)(40)/144$$
$$= 2200 + 14.5$$
$$= 2214.5 \, \text{psia}.$$

7. *Solution*:

$$P_{cr} = P_{cl}(\sigma_r / \sigma_l) = 18(24/72)$$
$$= 6 \, \text{psi}$$
$$\rho_{wr} = 62.4 \gamma_w / B_w = (62.4)(1.09)/1.02$$
$$= 66.67 \, \text{lb/ft}^3$$
$$\rho_{oST} = (62.4)(141.5)/(131.5 + API)$$
$$= (62.4)(141.5)/(131.5 + 35)$$
$$= 53.0 \, \text{lb/ft}^3$$
$$\rho_{or} = (\rho_{oST} + 13.56 \, gg \, R_s)/B_o$$
$$= [53.0 + (13.56)(0.7)(0.540)/1.24$$
$$= 46.87 \, \text{lb/ft}^3$$
$$P_c = h \Delta \rho / 144$$
$$h = 144 P_c / \Delta \rho = (144)(6)/(66.67 - 46.87)$$
$$= 43.6'.$$

NOTE: If surface densities are used (erroneously), $h = 57.6'$.

8. *Solution*:

$$P_c = 4.619 J(S_w)\sigma\sqrt{\frac{\phi}{k}}$$

$$P_c = 4.619 J(S_w) 27\sqrt{\frac{0.15}{130}} = 4.24 J(S_w).$$

S_w (%)	$J(S_w)$	P_c (psi)
22	1.5	6.36
25	1.2	5.09
30	1.0	4.24
50	0.66	2.80
80	0.38	1.61
100	0.36	1.53

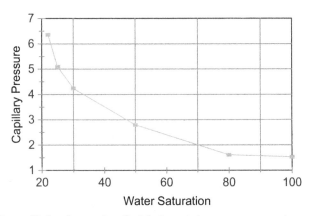

9. *Solution*: Calculate the fluid densities at reservoir conditions

$$\rho_{gr} = 13.56\, gg/B_g = (13.56)(0.75)/0.8$$
$$= 12.71\, \text{lb/ft}^3$$

$$\rho_{or} = (\rho_{oST} + 13.56\, gg\, R_s)/B_o = [52 + (13.56)(0.75)(0.7)]/1.20$$
$$= 49.26\, \text{lb/ft}^3$$

and then calculate the capillary pressure

$$P_c = H\Delta\rho/144 = (5)(49.26 - 12.71)/144$$
$$= 1.27\,\text{psi}.$$

Solutions — Chapter 2

1. *Solution*:

$$\frac{0.00113\,B_{o_j}}{\Delta y_j}\left[M_{o_{j+\frac{1}{2}}}\frac{P_{o_{j+1}} - P_{o_j} - \frac{\rho_o}{144}(h_{j+1} - h_j)}{\frac{\Delta y_{j+1} + \Delta y_j}{2}}\right.$$

$$\left. - M_{o_{j-\frac{1}{2}}}\frac{P_{o_j} - P_{o_{j-1}} - \frac{\rho_o}{144}(h_j - h_{j-1})}{\frac{\Delta y_j + \Delta y_{j-1}}{2}}\right] - \frac{q_o B_o}{V_b}$$

$$= \frac{S_o\,\phi\,c_o + S_o\,\phi\,c_f}{5.615}\frac{P_o^{t+\Delta t} - P_o^t}{\Delta t}.$$

2. *Solution*:

Iteration	z_{guess}	y	x	error	$z_{\text{calc.}}$
1	0	1.0	1.5	3.5	3.5
2	3.5	2.167	0.9167	−0.583	2.9167
3	2.9167	1.9722	1.0139	0.0972	3.0139
4	3.0139	2.0046	0.9977	−0.0162	2.9977
5	2.9977	1.9992	1.0004	0.0027	3.0004
6	3.0004	2.0001	0.9999	−0.0004	2.9999
7	2.9999	2.0000	1.0000	0.0001	3.0000.

$$1.\quad 2x + y = 4$$
$$y = 4 - 2x$$

$$2.\quad x + y + z = 6$$
$$x + (4 - 2x) + z = 6$$
$$-x + 4 + z = 6$$
$$-x + z = 2$$

3. $3y - z = 3$
   $y = (3 + z)/3$

4. $4 - 2x = (3 + z)/3$

5. $12 - 6x = 3 + z$
   $z = 9 - 6x$

6. $-x + z = 2$
   $-x + (9 - 6x) = 2$
   $-7x = -7$
   $x = 1$

7. $y = 4 - 2x$
   $y = 4 - (2)(1)$
   $y = 2$

8. $z = 3y - 3$
   $z = (3)(2) - 3$
   $z = 3.$

Matrix Solution Using Gaussian Elimination

For the three equations,

$$2x + y = 4$$
$$x + y + z = 6$$
$$3y - z = 3,$$

we can write the following matrix notation,

$$
\begin{bmatrix} 2 & 1 & 0 \\ 1 & 1 & 1 \\ 0 & 3 & -1 \end{bmatrix}
\begin{bmatrix} x \\ y \\ z \end{bmatrix} =
\begin{bmatrix} 4 \\ 6 \\ 3 \end{bmatrix}
$$

which is of the form

$$A \quad X \ = \ B.$$

Multiplying the second row by -2,

$$\begin{array}{ccccc}
2 & 1 & 0 & x & 4 \\
-2 & -2 & -2 & y & -12 \\
0 & 3 & -1 & z & 3
\end{array}$$

and adding the first row to the second row (the second row is the pivotal row; -2 is the pivot or number being eliminated),

$$\begin{array}{ccccc}
2 & 1 & 0 & x & 4 \\
0 & -1 & -2 & y & -8 \\
0 & 3 & -1 & z & 3.
\end{array}$$

Now, multiplying the second row by 3,

$$\begin{array}{ccccc}
2 & 1 & 0 & x & 4 \\
0 & -3 & -6 & y & -24 \\
0 & 3 & -1 & z & 3
\end{array}$$

and adding the second row to the third row (3 is the pivot),

$$\begin{array}{ccccc}
2 & 1 & 0 & x & 4 \\
0 & -3 & -6 & y & -24 \\
0 & 0 & -7 & z & -21.
\end{array}$$

At this point, we could use back-substitution and solve the bottom row as an equation, $-7z = -21$, etc. Our modified "A" matrix is an upper triangular matrix (note the zeroes in the lower left of the matrix). Using back-substitution at this point results in a Complete Gaussian Elimination.

We can also continue to simplify the "A" matrix to result in a unit matrix (all 1's on the diagonal) by dividing the third row by -7,

$$\begin{array}{ccccc}
2 & 1 & 0 & x & 4 \\
0 & -3 & -6 & y & -24 \\
0 & 0 & 1 & z & 3.
\end{array}$$

Multiplying the third row by 6 and adding it to the second row (and then dividing the third row again by 6),

$$
\begin{array}{cccccc}
2 & 1 & 0 & x & & 4 \\
0 & -3 & 0 & y & & -6 \\
0 & 0 & 1 & z & & 3.
\end{array}
$$

Dividing the second row by -3 and subtracting it from the first row,

$$
\begin{array}{cccccc}
2 & 0 & 0 & x & & 2 \\
0 & 1 & 0 & y & & 2 \\
0 & 0 & 1 & z & & 3.
\end{array}
$$

Finally, dividing the first row by 2,

$$
\begin{array}{cccccc}
1 & 0 & 0 & x & & 1 \\
0 & 1 & 0 & y & & 2 \\
0 & 0 & 1 & z & & 3
\end{array}
$$

and the resultant "A" matrix is called the unit matrix while our answers appear in the modified "B" column matrix.

NOTE: Do not use more than 4 iterations.

3. *Solution*:

Iteration	y_{guess}	x	z	error	$y_{\text{calc.}}$
1	0	2.000	-3.0000	7.0000	7.0000
2	7.0000	-1.5000	18.0000	-17.5	-10.5
3	-10.5	7.25	-34.5	43.75	33.25
4	33.25	-14.625	96.75	-109.375	-76.125

Effects of Weighting Schemes

New guess of y is weighted by $(0.5y_{\text{guess}} + 0.5y_{\text{calc.}})$

Iteration	y_{guess}	x	z	error	$y_{\text{calc.}}$
1	0	2.000	-3.000	7.0000	7.0000
2	3.5	0.25	7.5	-5.25	-1.75

3	0.875	1.5625	−0.375	3.9375	4.8125
4	2.84375	0.57813	5.53125	−2.95313	−0.10938
5	1.36719	1.31641	1.10156	2.21484	3.58203
10	2.15017	0.92492	3.45051	−0.52559	1.62458
15	1.96436	1.01782	2.89309	0.12473	2.08909
20	2.00846	0.99577	3.02537	−0.02960	1.97886
30	2.00048	0.99976	3.00143	−0.00167	1.99881

New guess of y is weighted by $(0.25y_{guess} + 0.75y_{calc.})$

Iteration	y_{guess}	x	z	error	$y_{calc.}$
1	0	2.000	−3.000	7.0000	7.0000
2	5.25	−0.625	12.750	−11.375	−6.125
3	−3.281	3.641	−12.844	18.484	15.203
4	10.582	−3.291	28.746	−30.037	−19.445
5	−11.946	7.973	−38.837	48.810	36.865
10	160	−78	477	−553	−393
11	−255	129	−767	899	644
12	419	−208	1255	−1460	−1041

New guess of y is weighted by $(0.75y_{guess} + 0.25y_{calc.})$

Iteration	y_{guess}	x	z	error	$y_{calc.}$
1	0	2.000	−3.000	7.0000	7.0000
2	1.75	1.125	2.25	0.875	2.625
3	1.96875	1.01563	2.906	0.109	2.07813
4	1.99609	1.00195	2.98828	0.014	2.00977
5	1.99951	1.00024	2.99854	0.002	2.00122
6	1.99994	1.00003	2.99982	2×10^{-4}	2.00015
10	2	1	3	5×10^{-8}	2
15	2	1	3	2×10^{-12}	2

Solutions to Problems

Iterative Solution Techniques
Successive Over-Relaxation

Using the iterative solution technique for approximating z, the SOR method can be applied if we update z using

$$z_{\text{new}} = z_{\text{guess}} + \Omega(z_{\text{calc.}} - z_{\text{guess}})$$

where Ω is the successive relaxation factor and z_{new} becomes our next guess. Note that when $\Omega = 1$, the SOR method is identical with our original iterative solution. For comparative purposes, an $|\text{error}| < 0.0001$ was used. When $\Omega > 1$, over-relaxation occurs, and when $\Omega < 1$, under-relaxation exists.

Ω	# of Iterations
0.5	13
0.6	10
0.7	8
0.8	5
0.85	4
0.9	5
1.0	7
1.1	10
1.2	13
1.3	17
1.4	24
1.5	38.

Iterative Solution Techniques
Matrix Solution Using Conjugate
Gradient-Like Method

For the three equations,

$$2x + y = 4$$
$$x + y + z = 6$$
$$3y - z = 3$$

we can write the following matrix notation,

$$
\begin{array}{ccc}
2 & 1 & 0 \\
1 & 1 & 1 \\
0 & 3 & -1
\end{array}
\quad
\begin{array}{c}
x \\
y \\
z
\end{array}
=
\begin{array}{c}
4 \\
6 \\
3
\end{array}
$$

which is of the form

$$ A \quad X = B $$

and set up the transpose of the A matrix by changing columns to rows

$$
A^{\mathrm{T}} =
\begin{array}{ccc}
2 & 1 & 0 \\
1 & 1 & 3 \\
0 & 1 & -1.
\end{array}
$$

Starting with an initial guess of

$$
\begin{array}{c}
x \\
y \\
z
\end{array}
=
\begin{array}{c}
0 \\
0 \\
0
\end{array}
$$

we can initially estimate the residuals by multiplying the original matrix A by the guess and subtracting it from the original column vector

$$
\begin{array}{c}
r1 \\
r2 \\
r3
\end{array}
=
\begin{array}{c}
4 \\
6 \\
3
\end{array}
-
\begin{array}{ccc}
2 & 1 & 0 \\
1 & 1 & 1 \\
0 & 3 & -1
\end{array}
*
\begin{array}{c}
0 \\
0 \\
0
\end{array}
=
\begin{array}{c}
4 \\
6 \\
3.
\end{array}
$$

Next we calculate the directional vector p by multiplying the transpose A^{T} by the residuals

$$
\begin{array}{c}
p1 \\
p2 \\
p3
\end{array}
=
\begin{array}{ccc}
2 & 1 & 0 \\
1 & 1 & 3 \\
0 & 1 & -1
\end{array}
*
\begin{array}{c}
4 \\
6 \\
3
\end{array}
=
\begin{array}{c}
8+6+0 \\
4+6+9 \\
0+6-3
\end{array}
=
\begin{array}{c}
14 \\
19 \\
3.
\end{array}
$$

And we will initially set an intermediate vector h equal to p

$$
\begin{array}{ccc}
h1 & & 14 \\
h2 & = & 19 \\
h3 & & 3
\end{array}
$$

We're now ready to perform the first iteration and can calculate the intermediate vector g, where $g = A * p$

$$
\begin{array}{ccccccccccc}
g1 & & 2 & 1 & 0 & & 14 & & 28 + 19 + 0 & & 47 \\
g2 & = & 1 & 1 & 1 & * & 19 & = & 14 + 19 + 3 & = & 36 \\
g3 & & 0 & 3 & -1 & & 3 & & 0 + 57 - 3 & & 54
\end{array}
$$

and calculating the coefficient a from $a = h^2 / g^2$

$$
a = \frac{14^2 + 19^2 + 3^2}{47^2 + 36^2 + 54^2} = \frac{566}{6421} = 0.0881.
$$

Now calculating our unknowns using $x^{\text{new}} = x^{\text{old}} + a * p$

$$
x^{\text{new}} = x^{\text{old}} + a \, p1 = 0 + 0.0881 * 14 = 1.23
$$
$$
y^{\text{new}} = y^{\text{old}} + a \, p2 = 0 + 0.0881 * 19 = 1.67
$$
$$
z^{\text{new}} = z^{\text{old}} + a \, p3 = 0 + 0.0881 * 3 = 0.264.
$$

Next, calculate new residuals from $r^{\text{new}} = r - a * g$

$$
r1^{\text{new}} = r1^{\text{old}} - a \, g1 = 4 - 0.0881 * 47 = -0.143
$$
$$
r2^{\text{new}} = r2^{\text{old}} - a \, g2 = 6 - 0.0881 * 36 = 2.826
$$
$$
r3^{\text{new}} = r3^{\text{old}} - a \, g3 = 3 - 0.0881 * 54 = -1.76
$$

and calculate the new h vector using $h = A^{\text{T}} r$

$$
\begin{array}{ccccccccc}
h1 & & 2 & 1 & 0 & & -0.143 & & 2.5407 \\
h2 & = & 1 & 1 & 3 & * & 2.826 & = & -2.5963 \\
h3 & & 0 & 1 & -1 & & -1.76 & & 4.5867.
\end{array}
$$

Determine coefficient b from $b = (h^{\text{new}})^2/(h^{\text{old}})^2$

$$b = \frac{2.5407^2 + (-2.5963)^2 + 4.5867^2}{14^2 + 19^2 + 3^2} = 0.06048$$

and finally by calculating p^{new} from $p^{\text{new}} = h^{\text{new}} + b\,p^{\text{old}}$, we have completed an iteration and are ready to start another one.

$$p1^{\text{new}} = 2.5407 + 0.06048 * 14 = 3.3875$$
$$p2^{\text{new}} = -2.5963 + 0.06048 * 19 = -1.4471$$
$$p3^{\text{new}} = 4.5867 + 0.06048 * 3 = 4.7681.$$

Following the same sequence for the second iteration, we will obtain

$$g1 = 5.3279$$
$$g2 = 6.7085$$
$$g3 = -9.1095$$
$$a = 0.2189231$$
$$x = 1.9757$$
$$y = 1.358$$
$$z = 1.3083$$
$$r1 = -1.3094$$
$$r2 = 1.358$$
$$r3 = 0.2343$$
$$h1 = -1.2607$$
$$h2 = 0.7515$$
$$h3 = 1.1237$$
$$b = 0.0998111$$
$$p1 = -0.9226$$
$$p2 = 0.6071$$
$$p3 = 1.5997$$

and finally, for the third iteration,

$$g1 = -1.2381$$
$$g2 = 1.2841$$
$$g3 = 0.22155$$

$$a = 1.0575442$$

$$x = 1$$
$$y = 2$$
$$z = 3$$

$$r1 = 0$$
$$r2 = 0$$
$$r3 = 0$$

and since the residuals are zero (or would be less than an acceptable tolerance in actual simulation), we have solved the system of equations.

Solutions — Chapter 3

1. *Solution*:

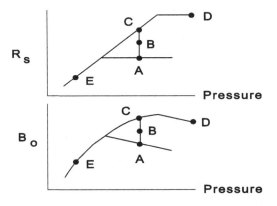

2. *Solution*: Calculate the spacing:

$$\frac{4000 - 400}{5 - 1} = 900 \,\text{psi increment}$$

resulting in the following equally-spaced table:

400
1300
2200
3100
4000 - bubble point
4900

This equally-spaced table is probably not appropriate because only the 4000, 4500 and 5000 psia values were used to determine the equally-spaced values at 4000 and 4900 psia.

Possibly okay since we need to span the pressure range (although you had certainly want more entries in your PVT table).

Solutions — Chapter 4

1. *Solution*: Calculate the normalized water saturations from

$$S_w^* = \frac{S_w - S_{wc}}{1 - S_{wc} - S_{orw}} = \frac{S_w - 0.104}{1 - 0.104 - 0.43}$$

S_w^*	k_{row}	k_{rw}
0	1	0
0.232	0.64	0.019
0.384	0.38	0.036
0.693	0.058	0.058
0.807	0.01	0.067
1	0	0.09

and recalculate the new water saturations by rearranging the normalized equation with the new end-point values

$$S_w = S_w^*(1 - S_{wc} - S_{orw}) + S_{wc} = S_w^*(1 - 0.15 - 0.40) + 0.15$$

S_w	k_{row}	k_{rw}
0.15	1	0
0.254	0.64	0.019
0.323	0.38	0.036
0.462	0.058	0.058
0.513	0.01	0.067
0.60	0	0.09

2. *Solution*: Calculate the normalized water saturations from (data shown for dataset 1)

$$S_w^* = \frac{S_w - S_{wc}}{1 - S_{wc} - S_{orw}} = \frac{S_w - 0.10}{1 - 0.10 - 0.28}$$

1.

S_w	k_{row}	k_{rw}	S_w^*
0.10	1	0	0
0.25	0.27	0.005	0.242
0.40	0.06	0.022	0.484
0.55	0.0043	0.080	0.726
0.60	0.001	0.125	0.806
0.72	0	0.35	1

2.

S_w	k_{row}	k_{rw}	S_w^*
0.20	1	0	0
0.30	0.26	0.007	0.278
0.50	0.06	0.035	0.556
0.60	0.013	0.092	0.741
0.70	0.01	0.23	0.926
0.74	0	0.35	1

3.

S_w	k_{row}	k_{rw}	S_w^*
0.30	1	0	0
0.40	0.45	0.0043	0.208
0.50	0.15	0.016	0.417
0.60	0.048	0.05	0.625
0.70	0.008	0.15	0.833
0.78	0	0.35	1

Next, at common relative permeability values, determine the appropriate value of S_w^* for each dataset and average the normalized water saturations

k_{row}	S_w^* (#1)	S_w^* (#2)	S_w^* (#3)	$S_{w\,\text{avg}}^*$
1	0	0	0	0
0.35	0.198	0.213	0.24	0.217
0.10	0.409	0.464	0.498	0.457
0.04	0.522	0.608	0.646	0.592
0.01	0.652	0.76	0.809	0.74
0.004	0.725	0.829	0.89	0.815
0.001	0.802	0.918	0.966	0.895

Calculate the average end-point values

$$S_{wc\,\text{avg}} = \frac{0.10 + 0.20 + 0.30}{3} = 0.20$$

$$S_{orw\,\text{avg}} = \frac{0.28 + 0.26 + 0.220}{3} = 0.253$$

and recalculate the new water saturations by rearranging the normalized equation with the average end-point values

$$S_w = S_w^*(1 - S_{wc} - S_{orw}) + S_{wc} = S_w^*(1 - 0.20 - 0.253) + 0.20$$

k_{row}	$S_{w\,avg}^*$	S_w
1	0	0.20
0.35	0.217	0.319
0.10	0.457	0.450
0.04	0.592	0.524
0.01	0.74	0.605
0.004	0.815	0.646
0.001	0.895	0.689

The same procedure is used to determine k_{rw} values.

3. *Solution*: k_{ro} values:

	Case 1	Case 2	Case 3
(Stone 1)	0.0998	0.0456	0.3422
(Stone 2)	−0.0350	−0.0750	0.2900

4. *Solution*:

S_g (%)	Elevation (ft ss)
75	−5420
50	−5424
25	−5426
0	−5428

$$\rho_{oST} = (62.4)(141.5)/(131.5 + API)$$
$$= (62.4)(141.5)/(131.5 + 45.1) = 50\,lb/ft^3$$
$$\rho_o = (\rho_{oST} + 13.56\,gg\,R_s)/B_o$$
$$= [50 + (13.56)(0.65)(0.480)]/1.18 = 45.96\,lb/ft^3$$
$$\rho_g = 13.56\,gg/B_g = (13.56)(0.65)/2.5 = 3.53\,lb/ft^3$$
$$P_c = H\Delta\rho/144$$

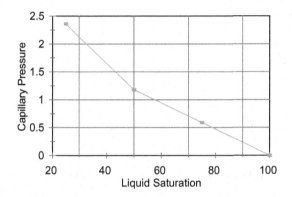

$S_g \%$	h (ft)	P_c (psi)
75	8	2.36
50	4	1.18
25	2	0.59
0	0	0

5. *Solution*: Reordering by decreasing (permeability/porosity) values since the porosity is not constant results in the following sequence of layer flood-out. Note that the sequence of layer flood-out is the same as for decreasing permeability ordering for this problem.

Layer	Permeability/Porosity (md)	Permeability (md)
5	937.5	150
3	533.3	80
1	222.2	40
4	125.0	10
2	35.7	5

$$1 - S_{orw} = 0.75$$

No layers flooded:

$$S_w = S_{wc} = 0.20$$
$$k_{rw} = 0$$
$$k_{row} = 0.85$$

Layer 5 flooded:

$$S_w = \frac{(0.2)[(4)(0.18) + (6)(0.14)+(5)(0.15) + (12)(0.08)]+(0.75)(0.16)(7)}{(4)(0.18)+(6)(0.14) + (5)(0.15) + (12)(0.08) + (7)(0.16)}$$
$$= 0.3403$$

$$k_{rw} = \frac{(0.4)(150)(7)}{(40)(4) + (5)(6) + (80)(5) + (10)(12) + (150)(7)} = 0.2386$$

$$k_{row} = \frac{(0.85)[(40)(4) + (5)(6) + (80)(5) + (10)(12)]}{(40)(4) + (5)(6) + (80)(5) + (10)(12) + (150)(7)} = 0.3429$$

Layers 5 and 3 flooded:

$$S_w = \frac{(0.2)[(4)(0.18)+(6)(0.14)+(12)(0.08)]+(0.75)[(0.16)(7)+(0.15)(5)]}{(4)(0.18) + (6)(0.14) + (5)(0.15) + (12)(0.08) + (7)(0.16)}$$
$$= 0.4343$$

$$k_{rw} = \frac{(0.4)[(150)(7) + (80)(5)]}{(40)(4) + (5)(6) + (80)(5) + (10)(12) + (150)(7)} = 0.3295$$

$$k_{row} = \frac{(0.85)[(40)(4) + (5)(6) + (10)(12)]}{(40)(4) + (5)(6) + (80)(5) + (10)(12) + (150)(7)} = 0.1497$$

Layers 5, 3 and 1 flooded:

$$S_w = \frac{(0.2)[(6)(0.14)+(12)(0.08)]+(0.75)[(0.16)(7)+(0.15)(5)+(0.16)(7)]}{(4)(0.18) + (6)(0.14) + (5)(0.15) + (12)(0.08) + (7)(0.16)}$$
$$= 0.5245$$

$$k_{rw} = \frac{(0.4)[(150)(7) + (80)(5) + (40)(4)]}{(40)(4) + (5)(6) + (80)(5) + (10)(12) + (150)(7)} = 0.3659$$

$$k_{row} = \frac{(0.85)[(5)(6) + (10)(12)]}{(40)(4) + (5)(6) + (80)(5) + (10)(12) + (150)(7)} = 0.0724$$

Layers 5, 3, 1 and 4 flooded:

$$S_w = \frac{(0.2)[(6)(0.14)]+(0.75)[(0.16)(7)+(0.15)(5)+(0.16)(7)+(0.08)(12)]}{(4)(0.18)+(6)(0.14)+(5)(0.15)+(12)(0.08)+(7)(0.16)}$$
$$= 0.6448$$

$$k_{rw} = \frac{(0.4)[(150)(7)+(80)(5)+(40)(4)+(10)(12)]}{(40)(4)+(5)(6)+(80)(5)+(10)(12)+(150)(7)} = 0.3932$$

$$k_{row} = \frac{(0.85)[(5)(6)]}{(40)(4)+(5)(6)+(80)(5)+(10)(12)+(150)(7)} = 0.0145$$

All layers flooded:

$$S_w = 1 - S_{orw} = 1 - 0.25 = 0.75$$
$$k_{rw} = 0.4$$
$$k_{row} = 0$$

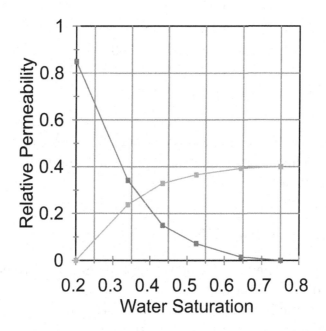

Final pseudorelative permeability values:

S_w	k_{rw}	k_{row}
0.20	0	0.85
0.3403	0.2386	0.3429
0.4343	0.3295	0.1497
0.5245	0.3659	0.0724
0.6448	0.3932	0.0145
0.75	0.4	0

Solutions — Chapter 5

1. *Solution*:

Location		Multipliers	
I	J	AX	AY
3	2	0	1
3	3	0	1
3	4	1	0
4	4	1	0
5	4	1	0
6	4	0	1

Note that the fault at the top of the grid requires no transmissibility modifications (transmissibility values on the boundaries are already zero).

Solutions — Chapter 6

1. *Solution*: The last row of aquifer cells may require pore volume modification for true aquifer extent. The grid selected will depend on the injection patterns to be studied. The grid shown was employed for a five-spot pattern created by converting existing producing wells.

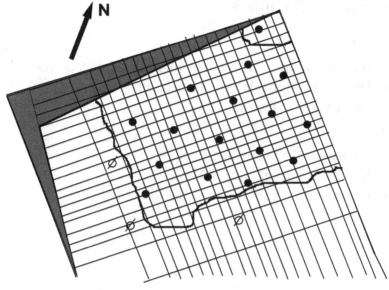

$$21 * 19 = 399 \text{ cells}$$

2. *Solution*:

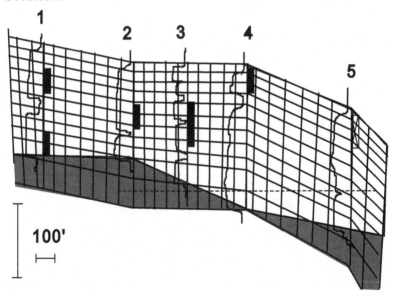

$$29 * 11 = 319 \text{ cells}$$

The shales would be represented by $AZ = 0$.

Solutions — Chapter 8

1. *Solution*:

   a. Increase aquifer size and/or increase aquifer permeability.

   b. Decrease water relative permeability (k_{rw}).

   c. Decrease skin factors.

2. *Solution*:

   a. Increase gas relative permeability (k_{rg}).

   b. Increase solution gas (R_s).

   c. Increase pore volume (or consider an aquifer).

INDEX

abandonment rate, 134
absolute permeability, 16, 143
adaptive implicit method (AIM), 64
advantages, 158
alternate diagonal, 70
Alternating Direction Implicit
procedure (ADI), 68
anisotropy, 19, 120, 133
aquifer, 122, 143
areal grids, 119
areal model, 4, 96, 104, 135
arithmetic average, 18
automatic history matching, 144
automatic timesteps, 136
automatic workover routine, 134
average pressures, 143
average the relative permeability sets,
91

black oil (or Beta) models, 8
bottom-hole pressure, 133
bubble point, 23, 25
bubble point pressure, 74

capillary pressure, 21, 22, 38, 85, 95,
104
cartesian (rectangular) coordinates, 4
cell property determination, 125
checkerboard, 71
compositional simulators, 8
computers, 156
coning, 105, 135, 152
coning models, 7, 151
Conjugate Gradient-Like (CGL), 70
conjugate gradient-like methods, 65
contouring, 126

corner point gridding, 129
critical (or connate) water saturation,
34
critical gas saturation, 38
critical water, 139
cross-sectional grids, 119
cross-sectional model, 5, 122
cross-sectional simulators, 104
cycle, 65

depths, 20
different relative permeability tables,
104
differential, 26
differential liberation, 24
direct solution, 65
directional permeability trends, 120
disadvantages, 159
drainage, 85
dual porosity, 154
dynamic pseudofunctions, 104

effective porosity, 13
efficiency factors, 134
elevations, 20
end-point scaling, 91, 105
equal-spacing of PVT data, 80
expansion of critical water, 105
explicit pressure solution, 59

fault simulation, 114
faults, 121
field study, 137
flash, 26
flash and differential, 83
flash liberation process, 24

Printed in the United States
By Bookmasters